COLORADO
AND THE
SILVER CRASH

COLORADO AND THE SILVER CRASH

THE PANIC OF 1893

JOHN F. STEINLE

THE History PRESS

Published by The History Press
Charleston, SC
www.historypress.com

First published 2021

Manufactured in the United States

ISBN 9781467147576

Library of Congress Control Number: 2021937192

To Mary, light of my life, this book is lovingly dedicated.

CONTENTS

ACKNOWLEDGEMENTS

It has been difficult to complete the amount of research needed for this book during the Covid-19 pandemic. Libraries have closed, travel has been disrupted and personal contacts have been minimized. Despite these handicaps, many people have assisted vitally in enabling me to complete this book.

The staff and volunteers of the Jefferson County Public Library system have heroically remained at their jobs, making sure that the libraries are run safely and that patrons can get the books they need. They deserve kudos along with other front-line workers who have been lauded during the pandemic. I would not have been able to acquire the research sources I needed without their help.

Jeanne Gripp, assistant to the director of the Cripple Creek District Museum, gave generously of her time and advice in choosing wonderful photos from the museum's collection during a delightful visit to that outstanding museum complex.

The staff members of the Colorado State Library who organized and created the Colorado Historical Newspapers Collection have built a resource that will remain essential for anyone studying the state's history. Kudos to them and to all the volunteers who have corrected the text!

At the Stephen H. Hart Research Center, History Colorado, Jori Johnson, collections access coordinator; Bethany Williams, collections access coordinator; Tamar McKee, manager of Stephen H. Hart Research Center; and Katie Bush, digital imaging studio manager, have all been tremendously kind and helpful in providing access to the photo and manuscript collections.

At Denver Public Library, Tom Williamson, cataloging assistant, Western History and Genealogy Department, has been instrumental in providing photos from the library's incredible collection.

Holly Henning, head of local history at the Lake County Public Library in Leadville, worked overtime to produce usable images from the library's Colorado Mountain History Photo Collection.

Lisa Keys, digital specialist for the Kansas State Historical Society, searched out an excellent image of Populist firebrand Mary Elizabeth Lease.

Miles Gallagher, curator of the National Mining Hall of Fame and Museum in Leadville, gave my wife and me a wonderful guided tour of this fantastic museum and much valuable background on the development of mining in Leadville. On his advice, we paid a revealing visit to the remaining mining infrastructure in California Gulch.

Voracious reader and expert proofreader Karen Meyer went through the manuscript with a very fine-tooth comb, pointing out my frequent mangling of grammar, spelling and punctuation.

The immense Library of Congress images collection presents a visual panorama of American history available to all without restrictions. This is one of the most valuable historical resources on the planet.

Finally, I would like to pay tribute to the old-time newspaper editors and reporters who offered the news of the nation and the world to their readers, along with a heaping dose of personal, salty, vivid commentary. They left no doubt as to where they stood on the issues of the day, whether Republican, Democrat, Populist or Feminist. Because of them, we can glimpse the personal lives of their readers—the true reality of history.

Introduction

TOP OF THE WORLD

The advertisement ran in the *Leadville Daily/Evening Chronicle* on April 20, 1893:

TABOR OPERA HOUSE
J.H. Cragg, Manager
Wednesday, April 26, '93
Engagement Extraordinary
The Champion of the World
JAMES J. CORBETT
Assisted by a select dramatic company, in
GENTLEMAN JACK
Superb Scenic Effects
The Campus at Payn College
Madison Square Roof Garden N.Y.
Training Quarters at Loch Arbour, N.J.
Olympic Club Arena, New Orleans
Admission $1.50
Gallery $1.00
Reserved seats now on sale at Newland's Corner Bookstore

The announcement caused quite a stir in Leadville: the heavyweight boxing champion of the world was about to arrive in the "Cloud City." Gentleman Jim Corbett himself! And so, on the evening when Corbett and his wife were to arrive—April 25, 1893—a "massive" crowd awaited the arrival of his train at the rail station. Speculation about his appearance must have rustled through the multitude. Surely the man who defeated the mighty John L. Sullivan—the "Boston Strong Boy"—must be a veritable giant!

The 1892 Sullivan-Corbett fight was already legendary among boxing devotees: how it had been the first heavyweight boxing match under the new Marquess of Queensberry rules, using boxing gloves instead of inflicting bare-knuckled mayhem; how Corbett's scientific style with its fancy footwork had allowed him to elude the hammer blows of the massive Sullivan; how an exhausted Sullivan had been knocked out in the twenty-first round; how Corbett had been hailed as inaugurating a new era in boxing, making it a respectable sport instead of an exercise in bloody, brutal hammering.

Corbett had been a bank clerk before turning to professional boxing. That and his penchant for natty, stylish clothes and his refined manners earned him the nickname "Gentleman Jim." His appearance fit perfectly with the men in Charles Dana Gibson's drawings of the idealized "Gibson girl" and her gentleman companion. The myth did not exactly fit with reality: John L. Sullivan, seen as a brutal streetfighter, was better educated than Corbett, having attended Boston College and having used boxing gloves in bouts as far back as 1880. But it was the myth that ruled on this April night in Leadville, Colorado.[1]

With all the anticipation of Corbett's arrival, according to the reporter for the *Herald Democrat*, the crowd failed to identify the world's champion:

> *A few seconds after the cars stopped a young person in silk hat stepped lightly to the ground and assisted a young lady to the platform. He wasn't a big man nor a stout man, especially, and looked more like a youthful graduate from a theological seminary than a man who had whipped Sullivan.... The large crowd in waiting hadn't the slightest idea that the medium-sized, pale-faced man with the slight stoop was the man they waited to see.*[2]

As the audience made its way to the Tabor Opera House the next evening, everything they passed on the way spoke of sound, secure prosperity. The buildings they saw had been constructed after the "silver boom" of the late 1870s had enriched Leadville and all of Colorado. Strongly built of stone and brick, they were silent testimony to the importance of Colorado's second-largest city, which many expected to become an even larger metropolis and the future state capital. Leadville's success was even more remarkable

Left: James J. "Gentleman Jim" Corbett (1866–1933) transformed boxing into the "sweet science" instead of a bare-knuckled brawl. *Library of Congress.*

Below: The wooden buildings of early Leadville in this photo were replaced during the 1880s and early 1890s by brick and stone edifices. *History Colorado.*

considering that it was in a remote, forbidding mountain location at an elevation of 10,115 feet.

Even the frame buildings displayed a certain distinction, such as the two-story home with its ornate front porch built shortly after the silver boom began by mining engineer August Meyer, who brought the scientific ore sampling process to Leadville. Temple Israel synagogue, built on land donated by "Silver King" Horace Tabor, showed architectural uniqueness with its twin spires. On the way to the theater, you might also pass the small frame home that Tabor and his now-divorced wife, Augusta, lived in before they struck it rich, a reminder of how quickly wealth could arrive, as if by magic, and how it could twist and destroy relationships.[3]

It would have been impossible not to notice the Roman Catholic Church of the Annunciation, built to accommodate Leadville's huge Irish population. Its tall spire with its intricate interior carpentry towered over every other building in Leadville. Some of the theatergoers may have been present at the church in 1886 when a young mining superintendent, James Joseph Brown, and his bride, Margaret Tobin Brown, were married there amid a gala assembly of relatives and friends. J.J. and "Maggie" Brown were to loom large in the legends of Leadville, Denver and the nation.[4]

Other churches marked Leadville's progression to respectability from an overcrowded, rowdy, violent, vice-ridden boomtown. The Presbyterians gathered in predestined order in their solid brick church with its open bell tower and colorful windows of hand-painted glass, one day to be known as the "Old Church." Episcopalians worshiped in rituals inherited from the Church of England at St. George's, illuminated by the light from windows with painted, furnace-fired glass.[5]

At the opposite end of the spectrum, the audience's route to the opera house would have shunned the more than three dozen houses of ill repute and the alleys with cribs where "customers" could obtain a cheap "poke." Prostitution was so solidly entrenched, despite being officially illegal, that an estimated 25 percent of the city's annual budget arose from fines levied on the madams and "soiled doves." The respectable folk attending the opera house performance might have shuddered if they happened to pass State Street with its evil reputation. The nadir of Leadville life and death was Stillborn Alley, a dirty sinkhole of theft, drug addiction, suicide, prostitution and murder. These poisonous streets and alleyways were reminders of a violent past that Leadville was trying to surmount, even if it included legendary characters such as Wyatt Earp, Bat Masterson, Tom Horn and Doc Holliday.[6]

State Street was Leadville's epicenter of vice. Gambling, prostitution, alcoholism and drug use infested its ramshackle buildings. *Denver Public Library, Western History Collection X-17656.*

West Fourth Street offered an entirely different picture; its elegant Victorian homes sheltering Leadville's silver moguls and their families earned the street its nickname, "Millionaire's Row." Especially notable was the long, one-story home built by Eugene Robitaille with its stained-glass window in the shape of a human eye. On West Ninth Street, another fashionable neighborhood, George E. King's two-story Second Empire–style house with its elaborate Mansard roof stood out.[7]

Proceeding down Harrison Street, the theater mavens would have passed one solid, reassuring institution after another. Smith's Dry Goods Store at Third and Harrison offered everything from blankets to underwear, rugs, patterns, hosiery and shirts, asserting that its "Prices Will Be Lower Than the Lowest Until the Clouds Roll By." If another reminder of the dominance of Leadville's "Silver King" was needed, the Tabor Grand Hotel with its imposing four-story height and elegant Mansard roof would have provided it. In the next block, the ruins of Joshua Fearnley's building, destroyed by fire, would soon be replaced by Fearnley's sprawling brick Iron Block, offering the possibility of even more commercial expansion. Then the American National Bank with its three-story red sandstone tower would have reminded

Left: George E. King's elaborate Second Empire–style house on West Ninth Street illustrates the eclectic nature of Leadville's architecture. *Photo by John Steinle.*

Right: Unlike the Tabor Grand, which has been converted to apartments, the Delaware Hotel still offers prime accommodations for travelers. *Photo by John Steinle.*

the throng of Leadville's economic expansion from a shabby collection of tents and shanties in the 1870s. The Delaware Hotel, with its elaborate façade, promised comfortable accommodations for Leadville visitors.[8]

A glance up at the hills surrounding Leadville brought a reminder of how the wealth that sustained the city was being created. The hills would have sparkled with light from candles, kerosene lamps and a few electric lights. All around the city, on Iron, Carbonate and Fryer Hills, stood the tall hoist houses, the machine sheds and outbuildings of the famed mines that made Leadville the "Carbonate City." Many of them in Leadville and throughout Colorado had names reflecting the Irish origins of their owners: the Maid of Erin, the Wolftone, the Robert Emmet, the Fenian Star, the Shamus O'Brien, the Irish Flag, the Shamrock and the O'Connell, among many others.[9]

Scattered among the hills were the small satellite communities that allowed miners to live closer to their workplaces, many of them separated from others by culture, nationality and language. Among them were Stumpftown, Strayhorse Gulch, Ibex Town, Chicken Hill and Evansville. Hispanic

Miners from around the world—Irish, Cornish, Italian, Slavic, Scandinavian, Hispanic and Black—were the backbone of Leadville's mining industry. *History Colorado.*

workers gathered in Stringtown, the Cornish "Cousin Jacks" in Jacktown, the Finns in Finntown, the Irish in Chicken Hill and African American families at the west end of State Street. Even the smell of the place indicated prosperity, with the stench from no fewer than fifteen ore smelters assailing the nostrils.[10]

Anticipation of a delightful evening would have grown as the crowd drew near to the opera house. Among the men, a glance right across the street at the Clipper Building would have wafted memories of bibulous evenings spent in the Board of Trade Saloon (still in operation as the Silver Dollar) with its ornate tile floor. In 1893, the Board of Trade was advertised as "The 'Old Reliable' and headquarters for all."[11]

Finally, when reaching the Tabor Opera House, the audience would have been escorted by ushers to their seats. They were described as a "large and fashionable" audience, "the extraordinary presence of ladies being noticeable." Corbett was received enthusiastically, the plot of the negligible play taking a back seat to his dashing appearance and charm comparable to those of Errol Flynn, who portrayed him in the film *Gentleman Jim*. He and

Left: Moyer Mine workers near Leadville load ore onto freight cars for transportation to local smelters, revealing the massive silver mining infrastructure. *Library of Congress.*

Below: The Arkansas Smelter was among the more than fifteen smelters processing the silver, gold, lead and copper ore that provided the basis for Leadville's wealth. *Library of Congress.*

The Tabor Opera House today, in Carol M. Highsmith's photo. The Opera House, now City of Leadville property, is being completely renovated. *Library of Congress.*

his wife left on the train the next morning for another theatrical appearance in Aspen at the Wheeler Opera House.[12]

The Tabor Opera House was itself symbolic of Leadville's prosperity, having been donated to the community by "Silver King" Horace Tabor. Built in 1879 at a cost to Tabor of $40,000, the theater boasted 450 red plush seats with another 400 of plainer design. The top floor originally held twenty-five rooms as an annex to the Clarendon Hotel. Most of the building material and furnishings had to be freighted to Leadville by wagon over treacherous mountain roads, since Leadville had no rail connection in 1879. Despite this handicap, the theater was completed in a record one hundred days.[13]

The Tabor couldn't compare to the famed opera houses of Europe (no Verdi or Wagner was ever performed there), and a clothing store and pharmacy occupied space in the building. However, it did provide at least a glimpse of culture and worldliness to the Leadville community beyond the everyday grimy routine of mining. The entertainment offered at the Tabor was lively and frothy, not the solemn and often tragic grand opera of the day. Such light operas as *Fra Diavolo*, *Martha* and *The Bohemian Girl* alternated with popular plays including *East Lynne*, *Oliver Twist* and *The Two Orphans*. Stark's Austro-Hungarian Orchestra and the Colorado University Glee and Banjo Club plus minstrel shows delighted audiences, while "Professor" William Winder revealed the wonders of phrenology, the "science of the mind."[14]

Many celebrities appeared at the Tabor before and after Corbett, including famed political cartoonist Thomas Nast and renowned Shakespearean actress Helena Modjeska. Probably the one most remembered today is Oscar Wilde. Wilde appeared at the Tabor in 1882, before his fame as a poet, playwright and novelist had fully developed. The packed audience at the opera house was surprised to see, instead of the spindly aesthete they expected, a tall, broad-shouldered man with, as the *Leadville Daily Herald* reported, "a stride more worthy of a giant backwoodsman than an aesthete." Attired in a dark velvet jacket and silk knee breeches and stockings, Wilde lectured for an hour and a half on "Art Decoration." At the end, the reporter

Original scenery from the Tabor Opera House stage was recently rediscovered and may be reused or displayed. *Denver Public Library, Western History Collection X-180.*

Oscar Wilde (1854–1900) lectured in Leadville in 1882 before gaining fame as a poet and playwright. His colorful tales of Leadville have become legendary. *Wikipedia*.

stated that he received scattered "applause similar to the desultory explosion of half a bunch of firecrackers." Many myths and legends surround Oscar Wilde's Leadville visit, some arising from the fertile brain of Wilde himself. Among them is his famous tale that he strolled into a "dancing saloon" with a sign over the piano: "Please do not shoot the pianist, he is doing his best."[15]

James J. Corbett was much more appreciatively received in Leadville, and he took the time to give a long interview with a *Leadville Herald Democrat* reporter, forecasting his upcoming fight with British battler Charles Mitchell and reminiscing about his championship bout with John L. Sullivan.[16]

As the "large and fashionable" audience left the Tabor Opera House on that charmed night in 1893, having enjoyed seeing "Gentleman Jim" in *Gentleman Jack*, everything they saw about them indicated wealth and prosperity. It seemed that Leadville would always be on an upward trajectory, on top of the world (almost literally) with the sky the limit. Few of them could have realized that the very element responsible for creating and sustaining Leadville would, within a few months, lead the city to disaster and despair.

Chapter 1

WINDFALL

Small quantities of gold along Colorado's Rocky Mountain Front Range were discovered in the fall of 1858 by a group of prospectors, including three brothers from Georgia, Levi, Oliver and William Greeneberry Russell. Though only a few hundred dollars in gold were panned from Cherry and Dry Creeks, rumor magnified it into a hoard of massive proportions. The result was the great Colorado Gold Rush. "Pike's Peak or Bust" was the cry in 1859 as tens of thousands made their arduous way across the Great Plains. They had a journey of well over five hundred miles to their El Dorado, fleeing the nationwide depression known as the Panic of 1857.[17]

The spirit of these "Argonauts" was captured in a poem published in the *Western Mountaineer* newspaper in Golden City, Colorado, in December 1859. It read, in part:

Full many thousands heard the sound
That heaps of shining gold were found
Close by
And every one thought in his spunk—
Why can't I find the biggest chunk?
I'll try![18]

If an Argonaut survived the trek to the "Pike's Peak Country," he would find chaos and confusion. Much of Colorado was still part of Kansas

Left: William Greeneberry Russell (1818–1877), along with twelve other men, sparked the great Colorado Gold Rush in 1858 with discoveries in Little Dry Creek. *History Colorado.*

Below: Early prospectors panned gold out of creeks and rivers to extract small amounts of gold dust and nuggets trapped in the gravel. *Library of Congress.*

Territory in 1859, but Kansas was consumed with the bitter struggle between pro- and antislavery factions known as "Bleeding Kansas." Little effort was being made by Kansas authorities to provide law and order in their distant Rocky Mountain region. The new arrivals reacted by meeting in October 1859 in Denver City and forming their own completely extralegal "government," unsanctioned by Kansas Territory or the United States government and dubbing it "Jefferson Territory."

Over 1858 and early 1859, a few small, straggling communities formed in Colorado, including Denver City, named after the Kansas territorial governor, James Denver. Amounts of gold filtered out of Colorado's creeks and rivers with pans, sluices and rockers were relatively small for each individual prospector. But over the winter of 1858–59, the gold rush entered a whole new phase. Experienced prospectors who had participated in previous gold rushes in Georgia and California—including John Gregory, William Greeneberry Russell and George Jackson—explored further into the mountains. They found lode gold, the source of the fine grains and nuggets of gold that filtered down into the streams and rivers below.[19]

The result was a rush into the mountains and valleys beyond. As the *Rocky Mountain News* later remarked, "Distance, danger, hardships, deprivations, are all forgotten in the wild search for gold, and the determination to dig for it."[20] Toll roads were chartered and built leading to the gold mining areas. Towns sprang up at the entrance to the canyons through which the toll roads ran, providing lodging, food and supplies for prospectors. Mount Vernon, Apex, Golden Gate City, Golden City, Boulder City and Bradford City were among these communities. Up in the rugged gulches and peaks of the Rockies, and in South Park, new towns formed around the mines, among them Gold Hill, Central City, Black Hawk, Idaho Springs, Breckenridge, Fairplay, Tarryall and Buckskin Joe.[21]

As prospectors searched deeper into the mountains, successive "stampedes" to supposedly richer mining regions came in waves. One such stampede occurred in 1859 when a group of fifty-six men and one woman known as the Plattsmouth Party searched for gold along Clear Creek beyond Idaho Springs. The three Griffith brothers from Kentucky, George, David and John, located gold in what became known as the Eureka and Griffith Leads and soon organized the Griffith Mining District. By June 1860, enough people had gathered to found Georgetown, presumably named after George Griffith. The *Rocky Mountain News* reported in May 1861: "I fully believe it to be the very richest of all mining districts yet discovered."[22]

Central City, Colorado, boasted of being central to the "richest square mile on earth" during the heyday of gold and silver mining. *Library of Congress.*

Another stampede centered on Ten Mile Creek and the Tenmile Range, about ten miles west of the town of Breckenridge. James McNulty and John Gibbs prospected for gold in the gulch later named after McNulty—and they found it. A true "stampede" resulted, with the rich placer deposits being mined by three thousand prospectors. By August 1860, the *Rocky Mountain News* was reporting, in a reference to the sluice boxes known as "Long Toms": "Several claims of astonishing richness have been opened in McNulty's gulch during the past two weeks. One of these claims has yielded in the eleven days ending August 14th over three thousand dollars to one 'elongated Thomas.'"[23]

Georgetown, Colorado, was dubbed the "Silver Queen" for its rich silver mines. By the 1890s, it was Colorado's third-largest city. *Library of Congress.*

Perhaps the most fateful stampede centered on a gulch located a few miles south of Ten Mile Creek on a branch of the Arkansas River. Prospectors searching the area in 1860 included Abe Lee. Running his fingers through his gold pan filled with dark sand, he found traces of gold and, according to legend, exclaimed, "Boys, I have all the gold of California in this pan!" Thereafter, the gulch was known as California Gulch, and the stampede was on! Within a few months, five thousand prospectors had swarmed California Gulch, even founding a small community they christened Oro City, using the Spanish word for gold.[24]

A third stampede led into the most remote, forbidding region in all of Colorado—the San Juan Mountains. Covering about twelve thousand square miles in southwestern Colorado, the San Juans were an almost impassable barrier to early prospectors. Yet the volcanic nature of much of the terrain had created enormous quantities of precious metals hidden in the forbidding heights. Eventually, more than five thousand square miles of

Prospectors progressed from panning to using wooden sluices known as "Long Toms" to process much larger quantities of gold-bearing gravel. *Library of Congress.*

the San Juans would be included in mining claims and $1 billion in silver and gold would be exported.[25]

The spring of 1860 saw a group of prospectors leave California Gulch, headed south to explore the San Juans. Led by Charles Baker, they would eventually announce that the southwestern mountains were immensely rich with mineral wealth. The *Rocky Mountain News* reported that they spent $3,000 to obtain a toll road charter from the territorial legislature and that they had purchased $2,000 in groceries to return to their mining claim. Yet a group of twenty-three experienced prospectors who traveled to the Baker diggings and returned to Denver reported, and the *Rocky Mountain News* blazed out in a headline, that "San Juan Is a Humbug!"[26]

Disappointed in their prospecting efforts, many men left "Baker's Park" in the spring of 1861 and investigated farther south along the Animas River. Enough people flocked to the area that they laid out the community of Animas City, the first town of that name in the area. In the nearby Tin Cup mining district, the prospectors in the Washington Gulch area, including Jim Taylor and Fred Lottis, found rich deposits of placer gold. Lottis was still living there, an eccentric bachelor, in 1881.[27]

In February 1861, Congress officially authorized creation of Colorado Territory, and William Gilpin was appointed the first territorial governor.

The *ad hoc* Jefferson Territory went out of existence, and boundaries for the territory, counties and mining districts were firmly established. What was also evident was that the territorial government was determined to keep Colorado in the Union as the Civil War enveloped the West.

By 1863, gold prospecting and mining in Colorado began to go into a slump. Some mining shafts were dug in early Colorado, but with miners lacking in labor and mining expertise, they tended to be too short to produce much gold. The Griffith brothers' mines near Georgetown were described as having shafts only twenty to thirty feet in length, while Charles Baker's "big" hole in the San Juans was said to be only five feet deep.[28]

Lack of the expertise needed to mine and process valuable ore was only one factor in the 1860s mining depression. The Civil War drew many miners away to join the Union or Confederate armed forces. Escalating conflicts with the Cheyenne, Arapahoe and other tribes left Denver and Colorado isolated as Native American warriors cut telegraph lines and attacked stagecoaches and wagon trains. The lack of railroad lines also slowed mining development. As if these elements were not enough, the Union Pacific transcontinental railroad route went through Wyoming, evading the towering, rugged Colorado mountains.[29]

Fortunately, by 1870, simultaneous events saved Colorado's economy from decline and revitalized the mining industry. The end of the Civil War meant that hundreds of thousands of young men discharged from the armies were looking for work and advancement. The continuing conflict between Native American tribes and federal troops moved elsewhere.

Efforts by Denver investors led by former territorial governor John Evans brought about construction of rail lines from Denver to the transcontinental road at Cheyenne and in Kansas—the Denver Pacific and Kansas Pacific. Professor Nathaniel P. Hill built Colorado's first scientifically designed chemical smelter in Black Hawk, opening a new phase in Colorado mining. Eastern capital, freed up by the end of the war, began to flow into Colorado as East Coast and English investors sank their assets into mining companies.[30]

Along with these developments, a new element began to gain importance in the overall Colorado mining industry: silver. In Colorado, the focus was on gold from the initial rush in 1858 through the late 1870s. There had always been some silver mining in Colorado, mostly centered on Georgetown and Silver Plume. But several factors in the late 1870s created a whole succession of silver booms.

The first was a change in national monetary policy relating to silver. Since the adoption of the U.S. Constitution, the country had followed a policy of "bimetallism"—minting both silver and gold coins and using both metals to support the nation's financial stability. The value of gold was set by the U.S. government in 1834 at $20.64 per ounce, while silver was valued at $1.29 per ounce, or a ratio of 16 to 1. Investors in silver or gold could bring ingots of those metals and sell them to a U.S. mint with no limit to the number of silver dollars that could be minted. This was known as "free coinage" of silver.[31]

International monetary policy tended toward adoption of a "gold standard," using only gold to support the soundness of a nation's economy and the value of the currency. The world's most powerful regime, the British Empire, had been on the gold standard since 1821, and the newly established German Empire went on the gold standard by 1873. Bowing to internal and international pressure, in that year Congress passed the Demonetization Act, banning the minting of silver coins (except for export) and free coinage of silver, effectively putting the United States on the gold standard. The price of silver dropped dramatically, and future silver champions referred to the act as the "Crime of '73." That same year, the drop in silver prices, overextension and stock manipulation by railroads and the resulting major bank failures caused the nationwide depression known as the Panic of 1873. During the next two years, 115 railroads went bankrupt and 18,000 businesses failed. The following years of economic instability are sometimes referred to as the "Long Depression."[32]

In order to combat the depression, aid the nation's farmers and help western mining investors, Congressman Richard P. Bland of Missouri and Iowa senator William B. Allison introduced a bill in Congress that would resume minting of silver dollars and require the U.S. Treasury to purchase silver at market value to the tune of no less than $2 million to no more than $4 million each month. The Bland-Allison Act was vetoed by President Rutherford B. Hayes, but Congress overrode his veto, and the law went into effect in 1878. The price of silver immediately rose, and this led to a new silver boom in Colorado. The interest in silver mining was also sparked by the renewed success of the Comstock Lode in Nevada, which produced $37 million in gold and silver in 1877, and the profitable Caribou mining district in Boulder County, Colorado.[33]

The intense search for new silver deposits led deep into the interior of Colorado's mountains. In the California Gulch area, prospectors had been bedeviled for years by the black gunk they called "black cement" fouling their sluices and rockers. Two of the prospectors, Alvinus B. Wood and William H. Stevens, decided to take some sludge and have it assayed by the Boston and Colorado Smelting Company. The results astonished them; the black material was granulated lead ore containing large amounts of silver. The two men immediately made a mining claim on Iron Hill. August Meyer, a German-trained mining engineer who ran an ore sampling facility for the St. Louis Smelting and Refining Company in Alma, heard of the new silver mine and came to look it over.[34]

Richard P. Bland (1835–1899) was one of free silver's strongest advocates and was considered the front-runner going into the 1896 Democratic Convention. *Library of Congress.*

Meyer began shipping ore from the Wood and Stevens mine back to St. Louis for refining in 1876. When he publicized the silver strike, thousands began struggling over the high mountain passes to the California Gulch area. Edwin Harrison, August Meyer and the St. Louis Smelting and Refining Company built an impressive smelter in California Gulch known as the Harrison Reduction Works. Over the winter of 1877–78, Harrison, Meyer, Alvinus Wood, local general store owner Horace Tabor and others organized the town of Leadville, named after the lead deposits that hid the region's riches for years. Tabor became the town's first postmaster. In August 1878, the *Colorado Transcript* reported, "Contrary to prophecy, Leadville is going ahead daily, with no indication of weakening. There is no let-up to the continued building and improvements of the city. There is no doubt but Leadville at present is by far the liveliest mining camp in the State, and probably in the whole continent." The new community grew with astonishing speed: by 1879, Leadville had an estimated twenty thousand residents, producing $11.5 million in lead and silver.[35]

The silver boom extended into the old Ten Mile mining district, about twelve miles north of Leadville, when George B. Robinson, owner of a Leadville general store, grubstaked two miners, Charles Jones and John Y. Sheddon, so they could explore for silver deposits. They found rich silver ore in the Ten Mile area, sparking yet another rush by miners into the new diggings by January 1879. The rush was so strong that multiple communities were founded near the booming silver mines, strung out along the Arkansas River: Robinson's Camp, Recen, Kokomo, Ten Mile City, Summit and Summit City were all founded in 1879, and a nearby smelter was built by William Graff. By 1881, the Ten Mile area produced $2.6 million in precious

FRANK LESLIE'S ILLUSTRATED NEWSPAPER

FRANK LESLIE'S ILLUSTRATED NEWSPAPER

Left: Prospectors explore for prospective riches amid the 1879 San Juan Mountains silver boom in this *Frank Leslie's Illustrated* engraving. *Library of Congress.*

Right: The enormous difficulty of transporting heavy mining machinery through the Colorado mountains is shown on this 1879 cover of *Frank Leslie's Illustrated. Library of Congress.*

metals, and Kokomo, the leading community, boasted of ten hotels, four "concert saloons" and a brass band.[36]

Even in the older communities of Georgetown and Silver Plume in Clear Creek County, silver mining redoubled, and Georgetown became known as Colorado's "Silver Queen." Georgetown grew to become the third-largest city in Colorado, after Denver and Leadville.[37]

The last silver boom, and the one facing the most difficulties and dangers, was the San Juan Mountain boom of the 1880s and early 1890s. There were two major issues: the opposition of the Ute people to invasion of their territory by prospectors and miners, and the difficulty of access to the extremely rough and mountainous region itself. There had been a wagon road into the San Juans as early as 1860, built by Ceran St. Vrain and his associates and running from Taos through Abiquiu to the mining areas. But

Ouray's 1886 Beaumont Hotel, seen in this Carol M. Highsmith photo, demonstrated the rapid development of the San Juan mining towns. *Library of Congress.*

prospectors coming from the north still had to follow narrow mountain paths using pack horses, mules or burros.[38]

Despite these handicaps, by 1870–71, several well-paying silver mines had been developed in the former Baker's Park area, and Major E.M. Hamilton freighted in the equipment to build a stamp mill. But further development could not progress without some formal agreement between the U.S. government and the various bands of the Ute people. They insisted on tribal rights to their land, guaranteed to them by treaty in 1868. In 1872–74, the Utes negotiated an agreement with government treaty commissioner Felix Brunot. After much hornswoggling and misrepresentation, they agreed to cede about 6.7 million acres for mining and settlement in return for a perpetual annual payment of $25,000. They retained the rest of their land under the 1868 treaty, and what remains of

Opposite: Chief Ouray (1833–1880) and his wife, Chipeta (1843–1924), led many Ute people in their struggle to survive invasion of their ancestral lands. *Library of Congress.*

Above: Cripple Creek miners of the 1890s starkly contrast today's miners—no hard hats, safety gear or electrical headlamps. *Cripple Creek District Museum.*

that area after further divisions is now the Southern Ute Reservation, with headquarters at Ignacio, Colorado.[39]

After this agreement with the Utes was signed, a tsunami of prospectors, miners, road builders and railroad workers swept over the San Juans, populating the towns of Silverton, Mineralville, Eureka and the second community named Animas City. In 1873, 1,500 mining claims were filed, and Silverton grew from a population of one hundred in 1875 to over five hundred in 1876. Lead, zinc and copper as well as gold and silver were shipped out of the San Juans. Ouray, named after the famous Ute chief, was incorporated in 1876.[40]

None of this massive development would have been possible without a fever of road building and railroad construction. Entrepreneur and Indian commissioner Otto Mears was responsible for much of it, earning the nickname "Pathfinder of the San Juans." Eventually, Mears would corral financing and organize construction of more than 450 miles of toll roads through the rugged San Juans, connecting the most important mines and communities.[41]

Even Mears's achievements in road building would be eclipsed by the arrival of the driving force in the economy, communications and technology—the railroad. Early Colorado rail development was hindered by lack of capital, the Civil War and the mining slump. But when the railroads finally started to build out from Denver in the 1870s, they helped spark tremendous economic development. W.A.H. Loveland's Colorado Central; former territorial governor John Evans's Denver, South Park and Pacific; and General William Jackson Palmer's Denver and Rio Grande all competed to be the first to reach the mining towns.[42] Building a road grade wide enough to run standard-gauge locomotives was almost impossible in the Colorado Rockies. Instead, most railroads used a narrow gauge of only three feet, six inches, compared with the standard gauge of four feet, eight and a half inches.[43]

By 1877, the Colorado Central reached Georgetown and Silver Plume, while the Denver, South Park and Pacific puffed into Leadville in 1880. The

Why railroads built in narrow-gauge configuration is revealed in this photo of a Denver and Rio Grande train rounding Marshall Pass. *Library of Congress.*

A Denver and Rio Grande work crew uses manpower and a steam shovel attempting to clear the track of twenty-foot-deep snow. *History Colorado.*

Denver and Rio Grande built deep into the San Juans. By 1887, General Palmer's railroad reached Durango, Ouray, Silverton and Aspen. Otto Mears jumped on the railroad bandwagon, organizing the Rio Grande Southern, Silverton Railroad and Silverton Northern. With the extension of the railroads, mining increased to such an extent that the Aspen area even surpassed Leadville in the value of ore production by 1891.[44]

Despite the encouragement given to silver mining by the Bland-Allison Act, by the late 1880s, the price of silver was dropping again because the massive amount of silver being mined was exceeding demand. There was enough concern that both Republicans and Democrats joined in forming "silver leagues" throughout Colorado; by 1892, there were nearly sixty silver clubs in the state. In Georgetown, "Buffalo Bill" Cody staged a "silver benefit" in 1886 at the McClellan Opera House featuring some of the performers from his Wild West extravaganza.[45]

Pressure from farming representatives and western mining interests led Senator John Sherman of Ohio to introduce a bill propping up the price of silver. The result, passed by Congress on July 14, 1890, required the U.S. Treasury to buy 4.5 million ounces of silver each month, replacing the

Bland-Allison Act and doubling the amount of silver being purchased by the U.S. government. This Sherman Silver Purchase Act was intended to boost silver prices, and it did temporarily: the price of silver shot up from $0.84 to $1.50 per ounce.[46]

But an unfortunate—even disastrous—provision in the Sherman Silver Purchase Act caused untold distress. According to the bill, investors could buy up silver and sell it to the U.S. Treasury, being paid in treasury notes that could be redeemed in either silver or gold. They would specify being paid in gold dollars, then sell the gold for more than they paid for the original silver. Speculators did this repeatedly, draining the treasury of its gold reserves. Thus, the act that was intended to benefit farmers and silver interests ended up being a recipe for disaster that would strike just as the silver industry of Colorado seemed to be on a solid, successful basis.[47]

Chapter 2

STORM CLOUDS

Despite the massive expansion of American industry, finance and technology during the 1880s, storm clouds appeared on the horizon in 1890. One of the striking factors in industrial and financial growth in the Gilded Age was the increasing interlocking of global finance. This had a huge influence on the developing financial crisis that led to the Panic of 1893 and the Silver Crash.

In 1890, the influential House of Baring Brothers bank in England faced disaster. The bank invested heavily in North America, making completion of the Canadian-Pacific and Atchison, Topeka and Santa Fe Railroads possible. Unfortunately, Barings also heavily invested in Argentinian government bonds. When political unrest and a failed wheat harvest forced Argentina to default, Barings was in danger of collapse. A consortium of the Bank of England and the national banks of France and Russia saved Barings. But the shock waves rippled out through all Central, South and North America as frightened British investors pulled back.[48]

The Barings Panic reached New York, where the brokerage firm of Decker, Howell & Co. went bankrupt. The firm's backer, the Bank of North America, was in danger of failing and was only rescued when J.P. Morgan assembled a consortium of New York banks to support it. Increasingly, jittery European investors began selling their American stocks and government bonds and demanding payment in specie, i.e., gold. Gold began flowing out of the United States Treasury.[49]

Tariffs were a volatile political issue throughout the nineteenth century, just as they have become today. Typically, agricultural regions felt victimized by high tariffs protecting American industries from foreign competition. Grover Cleveland's first inaugural address mainly expressed the Democratic Party's traditional opposition to high tariffs. The 1888 election of Benjamin Harrison as president and Republican majorities in Congress emboldened the Republicans to increase support for high protectionist tariffs. This issue again absorbed the country's attention in October 1890 when Congress passed a tariff bill promoted by Republican congressman William McKinley of Ohio. The new bill raised average tariff duties from 38 percent up to 49.5 percent. No duties were charged on sugar, molasses, coffee, tea or animal hides. It put a 70 percent duty on tin plate to promote the American tin plate industry. A young Democratic congressman from Nebraska, William Jennings Bryan, drew national attention for his outstanding oratory opposing passage of the tariff.[50]

The McKinley tariff raised prices on a wide range of products throughout the American economy. For the many working-class people and farmers already struggling through hard times, the effects of the tariff were yet another severe blow, adding to the root causes of the Panic of 1893. A *New York Times* headline and article of October 1890 gave eloquent expression to this situation:

> *UP GO THE PRICES NOW; HOW THE McKINLEY TARIFF TAXES THE NECESSARIES OF LIFE. MERCHANTS ARE MARKING UP ALMOST EVERYTHING THAT MEN WEAR, EAT, OR KEEP HOUSE WITH*
> *...WHO PAYS THE TARIFF TAXES?*
> *Let the facts, which are multiplying every day, tell who it is that pays the onerous tariff taxes....They will answer that the American people pay these taxes and that the burden of them rests most heavily upon the poor, inasmuch as there are very few of the necessities of life the prices of which are not increasing on account of the McKinley tariff.*[51]

Farmers were already in the throes of a depression in 1893. Wheat prices had dropped for years, and in 1892, they fell almost 20 percent. Overall, farm prices fell by two-fifths from 1870 to 1893. Farmers who borrowed money in better times buying land or investing in farm equipment found it impossible to climb out from under the burden of debt. In Washington State, 50 percent of the wheat crop was lost to heavy rains in 1893.

Left: Ohio congressman William McKinley (1843–1901) championed a protective tariff raising prices on imported goods, laying the groundwork for the Panic of 1893. *Library of Congress*.

Below: Endangered by economic forces beyond their control, farmers like these Nebraskans outside their sod house united to form the People's (Populist) Party. *Library of Congress*.

Midwestern, western and southern farmers felt they were being victimized on every side: by railroads, grain elevator companies, banks, mortgage brokers, commodities speculators, Wall Street, the national government, Great Britain and the weather.[52]

Contraction of the money supply also hurt. During the Civil War, the federal government used paper currency unbacked by gold or silver—"greenbacks"—to fund the Union war effort. Money was plentiful. But as greenbacks were withdrawn from circulation and the government stopped minting silver coinage, the money supply contracted. From a high of thirty-one dollars in 1866, the amount of money per person in the United States dropped to about twenty dollars by 1890. Farmers who bought land and equipment when money was available now found it difficult to pay debts when it was scarce.[53]

The year 1887 was an especially disastrous one for farmers and ranchers. Record high temperatures lasted for weeks across the whole country, while drought devastated farm crops. Huge wildfires ran rampant from Canada through the upper Midwest. An incredibly cold and bitter winter accompanied by deep snow killed most cattle in the Dakotas and Montana. Many ranches simply never recovered, including the Elkhorn Ranch owned by Theodore Roosevelt. President Grover Cleveland's veto of the Texas Seed Bill—designed to help destitute farmers in Texas, where 85 percent of the cattle in western Texas died—reinforced the attitude that the federal government cared nothing about the plight of ordinary working Americans.[54]

In December 1890, representatives from the Southern Farmers' Alliance, the Colored Farmers' Alliance, the National Farmers' Alliance and the Farmers' Mutual Benefit Association met in Ocala, Florida, and adopted what became known as the Ocala Demands. They demanded direct election of U.S. senators by the people instead of state legislatures, the breakup of monopolies, a graduated income tax, government ownership of all communications and transportation, a prohibition on ownership of land by noncitizens, limitation of national and state government expenditures to actual needs only, free coinage of silver and a return to bimetallism. This initial conference was followed by others in Cincinnati and Omaha, resulting in the formation of the People's Party or, as it was usually known, the Populist Party.[55]

Powerful leaders emerged, either belonging to the Populist movement or in alliance with it. Mary Lease crisscrossed the Midwest spreading the Populist message, becoming known as the "Kansas Pythoness," a reference to the ancient Greek Oracle of Delphi. Lease and her husband had failed at

Columbus, Nebraska, hosted this People's Party (Populist) nominating convention on July 15, 1890, at the height of Populist influence during the "Agrarian Revolt." *Wikipedia.*

homesteading, a failure she blamed on the railroads, banks and monopolies. Six feet tall, holder of a law degree and with a powerful voice, Mary Lease proclaimed, "The great common people of this country are slaves, and monopoly is the master!" Also attributed to her (probably apocryphally) is the cry, "Farmers should raise less corn and more hell!" Luna Kellie was another powerful Populist woman, secretary of the Nebraska Farmers' Alliance and a Populist newspaper editor and songwriter. Kellie's songs, such as "Dear Prairie Home," often lamented the plight of Nebraska homesteaders. The song mourns the death of a homesteader wife and loss of a Nebraska homestead to greedy creditors ("Shylocks"). One verse says,

Nebraska's fertile plain,
Where I toiled my manhood's strength away,
All that labor now is lost to me,
but it is Shylock's gain,
For that dear old home he claims today.

Luna Kellie's songs and the bitter experiences of many farmers explain why the greedy landlord is the standard villain in the melodramas of the day, coming to foreclose on the family farm and in the process trying to ravish the fair daughter of the household. The stalwart hero always arrives in the nick of time, shouting, "I'll pay the rent!" Unfortunately, no one arrived in time to save many midwestern farmers.[56]

Mary Elizabeth Lease (1853–1933) was one of the most prominent Populist orators and activists during the "Agrarian Revolt" of the 1880s–90s. *Kansas Historical Society*.

As these miners worked near Idaho Springs, Colorado, their future depended on the retention or repeal of the Sherman Silver Purchase Act. *Library of Congress.*

The Populist message found fertile ground in Colorado, among both silver miners and farmers on the state's eastern plains. What especially appealed to Coloradans was the Populist embrace of silver. Free coinage of silver, government support for high silver prices and bimetallism in support of the national currency would allow for expansion of the money supply. The resulting inflation would allow farmers to ask higher prices for their produce, workers to be paid higher wages and silver mines to charge more per ton for their ore. This would lead to easing of the debt burden on working-class citizens. It was felt that these factors would more than compensate for any increase in basic prices. The government specified the price of gold to be $20.67 per ounce. If it allowed free coinage of silver at a value of sixteen ounces of silver to one ounce of gold, this would officially fix the price of silver and stabilize its value at about $1.25 per ounce.[57]

Given the depressed situation of many farmers and the downward slide of silver prices, the Populists had a great opportunity to gain political clout in the 1892 elections. On their national ticket, they nominated James B.

Weaver for president and James H. Field for vice president. The Populist message resonated so well that Weaver and Field won 8.5 percent of the national vote and a majority in four western states, including Colorado. Weaver's political odyssey is typical of many Populist Party members. The luxuriantly bearded Weaver, an Iowan, had a sterling Civil War record, enlisting as a private and ending the war as a brevet brigadier general in the Union army. After the war, he supported the Republican Party, which had restored the Union, but he became concerned at the plight of Iowa farmers and joined the Greenback Party. After switching parties several times, he helped organize the fledgling Populist Party and rose through the ranks, as he had during the Civil War.[58]

In Colorado, Populists chose Davis Hanson Waite as their gubernatorial candidate. An attorney and newspaper editor from Aspen, his political journey was like Weaver's: a loyal Union Republican who became frustrated with conservative, pro-corporation Republican policies. Waite's campaign was aided by the fact that the silver issue divided the Democrats and Republicans. Both parties favored "sound money" and the gold standard. Many voters deserted the traditional parties to support Populist candidates who supported silver. Waite was elected Colorado's governor as a Populist in 1892, defeating his Democratic and Republican opponents.[59]

Another reason for the Populist triumph in Colorado was the party's support for women's suffrage. Many Coloradans saw women's suffrage and the silver issue as being united. It was a Populist, Representative J.T. Heath of Montrose and Delta County, who introduced the bill authorizing a referendum on women's suffrage to the Colorado legislature on January 16, 1893. The *Pagosa Springs News* stated, "The silver question and the suffrage question are vitally connected. A vote for equal suffrage is a vote for free silver. The man who goes to the polls on the 7th of next November, and votes to exclude the ballots of half the citizens of Colorado from the ballot box on account of sex, is a traitor to the white metal and to the best interests of the state."[60]

Just a month after the women's suffrage referendum bill was introduced in Colorado, the nation was fixated on a crisis that sent shock waves through the country's economy.

In 1893, the Philadelphia and Reading Railroad, one of the nation's oldest railways, seemed at a height of power and influence. The P&R leased the Lehigh Valley Railroad and the Central Railroad of New Jersey, making it one of the country's largest coal shippers. The P&R was attempting to buy the Old Colony Railroad, connecting much of New England, when the

Right: During the "Agrarian Revolt" in 1892, James Weaver (1833–1912) ran as a Populist Party candidate for president, winning five western states. *Library of Congress.*

Below: Locomotives such as Number 97 were idled when the Philadelphia and Reading Railroad suddenly went bankrupt, instigating the Panic of 1893. *Library of Congress.*

brakes went on and the metaphorical train came to a screeching halt. The company's aggressive president, Archibald Angus McLeod, had allowed the firm to acquire too much debt—$5 million from various banks. The creditors came calling, and on February 23, 1893, the Philadelphia and Reading went into receivership.[61]

The dismay from the failure of the P&R hardly subsided when the American economy was dealt another devastating blow. National Cordage Company, which had created a monopoly in making rope and

The bankruptcy of several influential companies sparked hysteria on the New York Stock Exchange, causing the Panic of 1893. *Wikipedia.*

cable, suddenly went into bankruptcy. National Cordage's directors spent heavily to corner the entire cordage industry in the United States while speculating deeply in the raw materials of ropemaking and issuing large stock dividends. This left the company without enough operating capital, and it folded on May 4, 1893. National Cordage stock dropped from thirty-nine dollars to twenty dollars, and this ignited a panic on Wall Street. General Electric and Union Pacific Railroad stock fell by about 7 to 10 percent in value. The stock market on May 5 closed on its most disastrous day until Black Thursday—October 24, 1929. The situation became more desperate when bank depositors, panicked by the Wall Street crash, began withdrawing their savings. Banks rapidly began to fail. The ripple effect extended to the South and West, where bank failures were common.[62]

Even the nation's great railroad systems, with their huge physical plants and thousands of employees, began to fail because of overexpansion on shaky financial grounds. The railroads built thousands of miles of unnecessary track just to forestall rival railroads from encroaching. During the 1880s, about seventy thousand miles of new track were laid. By July 25, 1893, the Erie Railroad failed and went into receivership. Eventually, the Union Pacific, Northern Pacific and Atchison, Topeka and Santa Fe would also cascade into bankruptcy. In total, sixty-five railroads would fold, and the effects would ripple throughout the economy, enveloping the iron and steel industries and throwing tens of thousands of workers out of a job.[63]

One of the most catastrophic effects of the Panic of 1893, and the one most directly affecting Colorado, was the collapse of silver. Before the stock market crash and the other catastrophes, silver had again been edging downward in price due to overproduction, even with the help of the Sherman Silver Purchase Act. Between 1892 and 1894, the Austro-Hungarian Empire, the Russian Empire, Chile and Persia all went on the gold standard. In May 1893, silver prices were down to eighty-three cents per ounce. Then a true hammer blow struck: the Herschell Committee of the British House of Commons recommended on June 26 that the Indian government stop the production of silver rupees and go on the gold standard like the rest of the British Empire. Immediately, the world price of silver dropped as the news sank in that demand from the Indian Mint would soon cease.[64]

President Grover Cleveland (1837–1908) believed that the Sherman Silver Purchase Act drained gold from the U.S. Treasury, and he led the repeal effort. *Library of Congress.*

When catastrophe strikes, the search for a scapegoat is inevitable. In the case of the Panic of 1893, the scapegoat was easy to find: silver. President Grover Cleveland and other political leaders pointed out that the Sherman Silver Purchase Act's provision allowing sales of silver to be paid by the government in certificates redeemable in gold was draining the nation's gold reserves. When U.S. Treasury officials announced on April 3, 1893, that gold reserves dipped below

$100 million, the commonly accepted lower limit, Cleveland and congressional leaders called for repeal of the Sherman Silver Purchase Act.

On August 7, 1893, President Cleveland called Congress into special session and said:

> *This matter rises above the plane of party politics. It vitally concerns every business and calling and enters every household in the land. There is one important aspect of the subject which especially should never be overlooked. At times like the present, when the evils of unsound finance threaten us, the speculator may anticipate a harvest gathered from the misfortune of others, the capitalist may protect himself by hoarding or may even find profit in the fluctuations of values; but the wage earner—the first to be injured by a depreciated currency and the last to receive the benefit of its correction—is practically defenseless. He relies for work upon the ventures of confident and contented capital. This failing him, his condition is without alleviation, for he can neither prey on the misfortunes of others nor hoard his labor.*[65]

Colorado politicians and newspapers leapt to the defense of silver mining and government purchase of silver. The *Leadville Daily/Evening Chronicle* trumpeted, "'No quarter, and no compromise' is the cry of the administration crowd. Let it be so. Let us have the one thing or the other. If the money value cannot be given to silver, with all of its ancient privileges, then let it be discarded altogether. We shall then see, and very soon, how the country can get along without it."[66]

Colorado's congressional delegation girded for an intense and prolonged battle over the silver issue. They were led by Senators Henry Teller and Edward Wolcott. Teller was a tall, lean Lincolnesque figure wearing the chin whiskers of a bygone time. He had served as Colorado's U.S. senator since Colorado became a state in 1876, with a short stint as President Chester A. Arthur's interior secretary. Teller moved to Colorado during the gold rush days in 1861 and as a Central City attorney was in the thick of many mining claim controversies. He was renowned as the greatest senatorial advocate of "free silver." Wolcott was a robust, imposing figure with a well-tended moustache and dark hair precisely parted in the middle. While his family was living in Ohio, Wolcott was an underage enlistee in the Union army during the Civil War at the age of sixteen. After graduating from Harvard after the war, Wolcott moved to Colorado, rising in status as an attorney, Colorado state senator and eventually U.S. senator in 1889. Like Teller, he

Left: Embroiled in fighting against repeal of the Sherman Silver Purchase Act, Colorado senator Edward Wolcott (1848–1905) participated in the forty-six-day filibuster delaying repeal. *History Colorado.*

Right: Senator Henry Teller (1830–1914) led senators battling repeal of the Sherman Silver Purchase Act, asserting that it was a moral crusade. *Library of Congress.*

was a tireless supporter of the silver cause. Wolcott was a jovial figure prized as both a host and a guest.[67]

Coloradans weren't the only defenders of silver in the Senate. Senators Jones and Stewart of Nevada, Dubois of Idaho, Pettigrew of South Dakota and Hansbrough of North Dakota were all "silver Republicans," and they were joined in battle as allies by Populists Allen of Nebraska, Kyle of South Dakota and Peffer and Martin of Kansas. They fought a stubborn and unrelenting battle on behalf of silver, using every tactic they could conceive to delay and derail the repeal effort.[68]

Public support for silver was mobilized on August 1, when a huge silver convention with more than one thousand delegates was held in Chicago. The *Leadville Daily/Evening Chronicle* reported that at the conference, an intense lobbying effort to Congress and labor organizations was planned: "It is to be a great moral crusade an awakening of the masses until the popular demand becomes so strong that the 'voice of the people will be stronger than the voice of gold' as the Populists put it."[69]

In this *Puck* magazine cartoon, Uncle Sam saves Columbia from drowning in silver coinage through repeal of the Sherman Silver Purchase Act. *Library of Congress.*

A bill to repeal the Sherman Silver Purchase Act was maneuvered through the House of Representatives, arriving in the Senate on August 27. The alliance of pro-silver senators filibustered the Senate for an astounding forty-six days, delaying a final repeal vote. Their hope was to wear down the "hard money" supporters and give time for public opinion to be heard on behalf of silver. The filibuster sessions were long and grueling. On October 11, the Senate met for thirty-eight hours, forty-five minutes. There were four roll call votes and thirty-nine quorum calls during that session. At one point, Senator William V. Allen of Nebraska held the floor for fourteen hours. On October 16, another marathon session lasted for eleven hours, with twelve roll call votes and thirteen quorum calls.[70]

Finally, on October 24, the filibuster ended when enough Democrats, pressured by the Cleveland administration, gave up on the silver crusade. On October 30, the repeal bill passed in the Senate. The *Idaho Springs News* lamented:

> *The long and gallant fight for silver on the floor of the United States Senate is at an end. The Republican silver senators have gone down like Custer's men on the Little Big Horn—overwhelmed by numbers, but without a thought of surrender. It has been a brave and manly fight that the Western republican senators have waged in the face of tremendous odds. Colorado's peerless leaders—Henry M. Teller and Edward O. Wolcott—have been constantly in the forefront of the fighting and have been tireless in their unselfish devotion to the cause.*[71]

Repeal of the Sherman Silver Purchase Act removed the last fragile barrier to disaster for the silver industry in Colorado. Now Coloradans already suffering through the Panic of 1893 grimly prepared themselves to withstand even harder times in the future.

Chapter 3

THE CRASH

The Panic of 1893 and the resulting Silver Crash devastated the Colorado silver mining industry, the railroads that connected with mining districts and all the businesses supplying both mines and railroads with equipment and supplies. In the aftermath of the Sherman Silver Purchase Act repeal, silver dropped to a low of $0.62¾ per ounce. All the communities that depended on silver mining to sustain them plunged into crisis. In fact, all the silver mining areas in the West, in Colorado, Nevada, Idaho and Arizona, shared the same fate.

Leadville was the Colorado city wedded most closely to silver mining. Silver was the reason for the city's very existence. When the full impact of the Panic of 1893 and the Silver Crash was starting to hit, the *Leadville Daily/Evening Chronicle* noted on June 29, 1893:

> *The situation in mining circles was practically the same as the day before, except that the screws were given another twist, and silver went down to 69, a break of 4 cents. One result of the conditions of affairs was the large number of idle men in the street; the avenue was as crowded during the afternoon and evening as it usually is on a Sunday.... The status of mining affairs is completely unsettled. Mining men are at a loss to know what to do. Everybody seems to be at sea, and the only thing that is being done is to watch and wait....Merchants have come to the conclusion that retrenching is the order of the day, and orders for many thousands worth of goods have been cancelled, as fast as the messages could be wired back to eastern houses.*

Without a railroad spur, mining supplies had to be transported by burro pack train, as in this photo from Ouray. *Library of Congress.*

Headlines in the Leadville *Herald Democrat* the very next day reflected the same situation:

THE MINES CLOSE DOWN
First Effect of the Tremendous Drop
In the Price of Silver Makes
Itself Manifest
Nearly Every large Mineral producer in
The Leadville District Closes Down—
Over a Thousand Men Now Idle and
Not a Pound of Ore Being Moved.

When the Silver Crash hit, mines and smelters like this one near Alma shut down overnight. *Library of Congress.*

The full impact of the Panic of 1893 and the resulting Silver Crash hit Leadville like a mine cave-in. Ninety mines completely shut down, at least twenty businesses failed and roughly 2,400 men were laid off—90 percent of the available workforce. Many men were seen buying railroad tickets to leave town and seek work elsewhere.[72]

The same conditions existed throughout Colorado's silver mining regions. The *Denver Republican* reported on mining near Creede on June 30, 1893: "All the producing camps practically have stopped shipping, the owners preferring to let the silver lie in the ground rather than sell it at the present low price. The Amethyst was the first to reduce force and did so early in the week. The New York and Chance and Bachelor will ship no more after tomorrow unless the price goes up." In Georgetown, forty-five mines closed, leaving six hundred men unemployed. The railroads cut their services to the smelters at Denver, Pueblo and Leadville. The *Engineering and Mining Journal* reported on September 16, 1893, "The southwestern part of the state—the 'Silvery San Juans'—is still in a woeful condition, owing to the closing down of the mines." In Pueblo, all the smelter operators stopped buying ore by July 6, and fifteen to twenty of the most important merchants cancelled orders for fall goods.[73]

On July 1, the *Greeley Tribune* reported that the most important Colorado mine owners and managers met at the Brown Palace Hotel in Denver and decided to stop all mining and smelter operations. The *Tribune* noted, "The mines cannot be operated at the present price and to stop further loss all operations will come to a standstill and 25,000 to 30,000 men will be thrown out of employment." David Moffat was one of the largest mine owners in Colorado, as well as being deeply involved in railroad investment and management. In an interview with the Associated Press, he remarked, "In the mines I control there are fully two thousand men employed and those people furnish food for fully five times that number. Ten thousand souls outside of this are employees of the railroads, of the smelters, of the oil wells, of the coal mines, and the multitude of middlemen generally who are all directly affected in the working of the mines."[74]

It was inevitable that those who were suffering through the Silver Crash would look for scapegoats. To some, the fault lay with President Grover Cleveland, and the *Aspen Daily Times* exhorted its readers to "go to church today and pray that the devil takes Cleveland before night." The Silver Plume *Silver Standard* exclaimed, "Free coinage of silver was discontinued by means of fraud and bribery." Others blamed English investors and inevitably fell back on chauvinism and religious prejudice. The *Salida Mail* quoted a poem written by a Leadville miner that went, in part,

But with damnable purpose and selfish view,
Came the English lord and the Wall Street Jew,
And to ruin of many and the benefit of few,
Our free-coinage laws they did undo—
Listen to my tale of woe.
Chorus—Hard trials—times look blue;
Silver and lead not worth a sou,
All account of the goldbug crew—
Listen to my tale of woe.[75]

Bedraggled miners' children in Leadville play with their burros amid the desolate industrial landscape of a Colorado mining town. *Library of Congress.*

For many Coloradans, the enemy was the "money interest," the combination of Wall Street stock manipulators, greedy "plutocrats," dominant national corporations, railroad presidents and "goldbug" advocates of the gold standard. The Carbondale *Avalanche* urged Colorado railroads to unite in pressuring the government to restore silver to its previous status: "In this emergency every influence has its might, and Colorado railroads cannot afford to trust all to the weak endeavors of a few silver miners, any more than the farmers of the west and south can afford to sleep until after the battle is lost."[76]

The lives of Colorado miners were tough enough without the additional trauma of the Silver Crash. They led lives of backbreaking labor almost inconceivable to the people of today. Most miners were paid about three dollars a day, usually for an eight- to ten-hour shift. The work week was six

Left: This Colorado miner is "single-jacking," drilling a hole to contain explosives that will break up ore-bearing rock. *Library of Congress.*

Opposite: Early pneumatic drills were called "widowmakers" because of the silicosis-causing clouds of rock dust they stirred up. *Cripple Creek District Museum.*

days, with the only off day being Sunday. Miners went to work in the early morning dark, worked in oppressive darkness and often went home in the dark. Working conditions in the mines were extremely dangerous, with the possibility of cave-ins, death from mistimed explosive charges and poisonous mine gas ever-present. One of the worst Colorado mining accidents happened right at the beginning of 1893, when twenty-four coal miners were killed by a gas and dust explosion in the Como No. 5 mine, owned by the Park Coal Company. On average, about three thousand miners died each year across the United States in mining accidents. Work inside the mine was only lit by candles at the time of the Silver Crash, with the first carbide mining lamps being introduced several years later.[77]

Miners could work alone "single-jacking." The miner would pound away at a drill using a four-pound sledgehammer. When the hole was deep enough, it would be packed with explosives and the detonation would break up the rock so it could be loaded onto oar carts. Or they could work in a team "double-jacking," with one miner holding the drill while another swung an eight-pound sledgehammer to hit it.

In addition to quick death from accidents, miners faced slow death from silicosis. The disease is caused by inhalation of dust from hard rock and minerals, grinding away the walls of the lungs. Ironically, an advance in

technology worsened the possibility of death from silicosis. By about 1880, mine owners began introducing pneumatic drills. With these, one man could drill more quickly than six men single-jacking. But the pneumatic drill kicked up a huge cloud of rock dust, which was inhaled by the drill operator and the men around him. Many died from silicosis. Pneumatic drills became known as "widowmakers" as a result.[78]

By 1893, most mining communities were past the "tent and shack" phase, though miners' housing was still primitive. Anne Ellis described the interior of a typical miner's cabin in Bonanza, Colorado:

> *The usual miner's cabin—a bunk built in one corner, a mattress of straw or pine boughs and over this blankets, with a top cover of canvas, and a pillow with a dark calico slip (quite shiny in the middle). At the head of this bed is a shelf, either nailed, or on pegs set in the logs; on this shelf are matches, a candle, pipes and tobacco, shaving mug and razor, and a small box holding thread, needles, and buttons. Against the wall is a home-made table, on which are cans containing sugar and salt; also a can of condensed milk, a few tin plates and cups, these being turned upside down when not in use, to keep mice and dirt out. Over the dishes, knives and forks, the dish-pan was turned; on top of and covering all this was spread the not very white dish-towel, made from an old flour sack. The chairs were blocks of wood, or two pieces of board nailed on a slant, with a seat fastened on; the stove was a tiny sheet-iron affair, with a coffee-pot on the back. Under a curtain in the corner hung their "other clothes"; near the door, on a box turned on end, would be a water-bucket and wash-pan.*[79]

Women in mining communities worked harder than the men in many ways. Bearing and raising the children; sewing, washing and mending clothing; keeping the house clean; cooking the meals and washing dishes; planting and tending gardens; managing cows and chickens; often homeschooling the children; taking in extra sewing or washing to make ends meet; doctoring neighbors when they were sick; enduring endless anxiety about the safety of their menfolk in the mines—all these burdens made the women of mining families old before their time. Anne Ellis recalled her mother's life in the tiny mining town of Exchequer:

> *Evenings, when the day's work is done, the cows in and milked, wood off the hill and chopped, water brought from the creek, the chicken-house fastened (Mama doing all this, with the help of some of the children), Henry in his room; the rest of us gather round the heating-stove, Mama always with sewing and mending. She made all our clothes, dreading the boys' suits, making underwear from flour sacks—never enough to go around. Besides this, she would hook rugs and piece quilts. She would often drop her work*

In smaller mining communities and remote mines, tiny log cabins were typical homes. These miners pose with their prized and beloved burros. *Library of Congress.*

> *and cry and when asked the reason would say, "Just blues." I wish now that I had read to her, but I was young and selfish, only reading to amuse myself. There was one, Jose, who would slip to her side, lean on her, and cry with her. There was never any affection shown, except that I have seen her love the latest baby, or gather the one next to it in her arms, deploring the fact that he had to be weaned, "and only a little un yoreself."*[80]

Food seemed to be plentiful: the Miner's Meat Market advertised in Ouray's *Solid Muldoon* newspaper in 1892 offering "Fresh and Cured Meats, Poultry, Fish, Veal and Sausage Vegetables of all kinds to suit purchasers." Harriet Backus remembered eating "Maryland oysters shipped frozen in cans, canned salmon, chicken, and turkey which I kept on hand. George had a ravenous appetite requiring plenty of bacon at breakfast, an abundance of meat in his lunch bucket and for dinner.…We bought fruit, vegetables,

Left: This old miner's home in Telluride, Colorado, is typical of the rustic housing lived in by workers in the larger mining communities. *Library of Congress*.

Below: The Silver Moon Restaurant in Victor, probably owned by Mrs. Swanson (*second from right*), was one of many Colorado female-owned businesses. *Cripple Creek District Museum*.

Denizens of this Telluride saloon could enjoy gambling away their hard-earned cash at the roulette wheel and squandering their pay on liquor. *Library of Congress.*

and milk by the case. Occasionally two or three neighbors divided a case of something special to vary the monotony without investing too much. Canned food and eggs were allowed to freeze but never potatoes or oil."[81]

With very limited leisure time or spare money, miners often hit a saloon after the day's work was done. Creede, the newest and rawest of the mining towns, was especially notorious for its forty saloons, gambling dens and brothels roaring day and night. Silverton matched Creede in its array of saloons, gambling houses, opium dens and brothels. In Ouray, it was said that as many as one hundred women were employed in local brothels, and the numerous saloons included the Bird Cage, Temple of Music, Bon Ton, Clipper and Gold Belt. Alcoholism was as great a scourge in nineteenth-century America as the opioid crisis is today.[82]

With men outnumbering women by as much as twenty to one in some mining communities, prostitution flourished. Prostitution in mining towns is often treated as a joke, with ladies in Victorian scanties being pushed in the annual Madame Lou Bunch bed race in Central City. But for the many women forced into a life of prostitution by adverse circumstances, it was no joke. Death from murder, disease and suicide was common. Even Cripple

Creek's most prosperous madam, Pearl DeVere, died from an overdose of morphine in 1897. Her elegant house of ill repute, the Old Homestead, is preserved as a museum. Though some madams prospered and became somewhat respected in their communities, the ordinary "working girls" faced a tough, often short life.[83]

At least in the older, more settled mining communities, life was enriched by much more than alcohol and illicit sex. Every town had its baseball team. The Aspen team was good enough to not only play Colorado clubs but also to tour Utah in 1895. When they played Glenwood Springs, the town band led a parade of the team through the streets of Aspen, while the Glenwood Springs town band joined in to play throughout the game.[84]

Both men and women enjoyed the fellowship of all the various fraternal organizations that had enormous memberships in nineteenth-century America. Every city and town had its lodges of the Knights of Pythias, Odd Fellows, Ancient Order of United Workmen, Fraternal Order of Masons,

Local baseball teams like this Cripple Creek team were objects of fierce community pride, offering relaxation away from grueling mining work. *Cripple Creek District Museum.*

Grand Army of the Republic, United Confederate Veterans, Elks, Knights of Columbus, Ancient Order of Hibernians, Turnverein, Independent Order of Good Templars, American Woodmen and many others. Women formed their own groups, such as the unique Social Ethics Women's Club in Idaho Springs, or joined fraternal auxiliaries such as the Masonic Order of the Eastern Star or the Woodmen of the World Royal Neighbors of America. Practically speaking, it was beneficial to join because many of these groups offered insurance benefits covering injury, death and burial expenses. In an age before unemployment, injury and disability coverage, these benefits were of enormous value to miners. In her book *The Life of an Ordinary Woman*, the extraordinary Anne Ellis recounted how the Knights of Pythias took care of all funeral arrangements when her husband, George, was killed in an accidental mine explosion.[85]

The lives of Colorado's people in the late 1800s were filled with music. Just a short survey of newspapers from that era reveals music being played by town bands, church choirs, small orchestras and soloists for nearly every public occasion such as Decoration Day (now known as Memorial Day), the Fourth of July, baseball games, funeral processions, commencements, Christmas and other religious holidays, balls and dances and Robert Burns

The Leadville Drum Corps with its colorful Zouave uniforms was typical of the many musical groups that enlivened miners' lives in the 1890s. *History Colorado.*

Night celebrations. Americans of that time loved to dance, and there was no "macho" attitude toward it among men. Waltzes, square dances, polkas, galops, schottisches and quadrilles were among the popular dances of the day.[86]

Despite the diversions available to miners, there was no escaping the Silver Crash. Thousands of Coloradans were unemployed and desperate. Throughout the summer and fall of 1893, the suffering increased, with Leadville being especially hard hit. With 90 percent of the workforce unemployed and hundreds leaving town, it seemed that Leadville was doomed to become a ghost town.

Ironically, it was Leadville that recovered most quickly of all the Colorado silver mining communities, because a group of determined mine owners and one daring engineer saw opportunity when others were stunned and unable to react. In 1891, a group of Leadville men formed a new corporation called the Ibex Mining Company. The directors included John "Leadville Johnny" Campion, August Meyer, Max Boehmer, A.V. Hunter and William Byrd Page. Among the employees and stockholders was a young mining engineer, J.J. "Jim" Brown. The new company purchased ownership of several mines on Breece Hill. By 1893, J.J. Brown was the superintendent of all the Ibex properties. He and the directors, with their knowledge of local geology, felt that there was a possibility of uncovering gold veins farther back in the silver mines. Brown began exploration in the Little Jonny mine but was flummoxed by the constantly flowing loose sand around the mine shaft. Using a system of timbers supporting tightly packed hay bales, he was able to overcome this problem and uncovered incredibly rich veins of gold and copper. This discovery in the Little Jonny sparked the resurrection of Leadville.

By November 1893, the Leadville *Herald Democrat* trumpeted:

> *GREATEST ON EARTH*
> *The Little Johnnie Comes to the Front*
> *As the Banner Gold Producer*
> *Of the Country*
> *…The Ibex Mining Company, operating the Little Johnnie mine, is shipping one hundred and fifty tons a day of ore carrying over twenty-five dollars worth of gold to the ton. This is a value yield of $3,750 a day, or $1,868,750 a year, which is believed to be a greater yield than that of any other gold property in the United States.*

Mining supervisor J.J. Brown, with his son Lawrence "Larry" Brown, became a millionaire through skillfully managing Ibex Company mines. *Denver Public Library, Western History Collection X-21669.*

In the aftermath of the gold strike, J.J. Brown was granted 12,500 shares of Ibex stock and a seat on the board of directors, making him a millionaire and one of the most influential people in Leadville.[87]

The success of the Little Jonny gave additional impetus to a second gold rush throughout the southern Colorado "gold belt." Already, in October 1890, a local cowboy and prospector named Bob Womack uncovered a vein of gold near a waterway known as Cripple Creek. Another incredibly rich vein of gold, known as the Independence Lode, was discovered by Winfield Scott Stratton. Soon a tiny community of five hundred people mushroomed to ten thousand. Older gold mining communities like Central City, Black Hawk and Idaho Springs gained new life as the United States went on the gold standard and demand for gold increased. In Jefferson County, several new communities were founded, centered on smaller lodes of the yellow metal.[88]

Leadville community leaders sought a way to announce the town's recovery to the world. The answer was to turn the freezing winter weather to Leadville's advantage and construct a marvelous Ice Palace. Although ice castles had previously been built in Montreal and St. Paul, the Leadville Ice Palace was the most wondrous of all, covering 10,000 feet of floor space enclosing a skating rink, a theater, two ballrooms, a banquet room

Leadville's Little Jonny Mine sparked a mining renaissance when massive quantities of gold and copper were discovered deep within it in 1893. *History Colorado.*

Leadville initiated construction of an elaborate Ice Palace, opening in January 1896. The palace drew 250,000 visitors before closing in May. *Library of Congress.*

and an exhibit area, all lit by one hundred arc lights and one thousand incandescent bulbs of various colors. A merry-go-round was situated outside, and a toboggan slide to the Ice Palace was 1,800 feet long. Work began on November 30, 1895, but warm weather delayed completion. The grand opening happened on January 4, 1896. The Ice Palace hosted an estimated 250,000 visitors before most of it closed in late March, though the skating rink remained open several months more. It was not the financial bonanza that Leadville leaders expected, and Leadville never created another Ice Palace.[89]

The Festival of Mountain and Plain in Denver and the Ice Palace in Leadville only camouflaged the reality of the Silver Crash. Financial disaster still stalked Colorado, even affecting those who had prospered most during the silver boom.

Otto Mears was one of those who faced disaster during the Silver Crash. Despite all his efforts through the years in bringing transportation routes to the San Juan Mountains, he found himself in serious financial difficulties in 1893. As recently as January, the *Pagosa Springs News* reported that Mears's Rio Grande Southern would build a railway route to the new southern gold mining regions. Mears even planned to extend the Rio Grande Southern south to Phoenix and from there to the West Coast. But as freight and passenger traffic in the San Juans shut down during the Silver Crash, the Rio Grande Southern hemorrhaged money. The bank on which Mears depended for financial backing went under. By August, the Rio Grande Southern was in receivership, and Mears lost control of the railroad to the much larger Denver and Rio Grande.[90]

The whole rise and fall of silver in Colorado is exemplified by one man: Horace Austin Warner "Haw" Tabor. Tabor and his wife, Augusta, moved to Colorado from Kansas in 1859, prospecting and tending general stores in different areas. They ended up in Leadville, where Tabor helped found the city and became the postmaster and mayor. With incredible luck, Horace grubstaked two miners with the promise of one-third of the proceeds if they struck it rich. They did: their Little Pittsburg mine produced huge quantities of silver ore. Tabor eventually sold his partnership in the Little Pittsburg for $1 million and then bought the even more productive Matchless and Chrysolite mines. Tabor used his wealth to benefit Colorado, funding construction of the Tabor Opera House and Tabor Grand Hotel in Leadville and donating the land on which the Temple Israel synagogue was built. In Denver, he constructed the Tabor Grand Opera House, said to be Colorado's most beautiful theater, and the mercantile complex known as the Tabor Block.[91]

Left: Elizabeth "Baby Doe" Tabor (1854–1935) married Horace Tabor after his scandalous divorce but shared his financial ruin. *Denver Public Library, Western History Collection Z-14609.*

Right: Horace Tabor (1830–1899) became Colorado's "Silver King" in the 1870s, but the 1893 Silver Crash destroyed his struggling financial empire. *Library of Congress.*

Tabor's wealth, influence and popularity won him election as lieutenant governor of Colorado in 1878, serving for six years. When Senator Henry Teller resigned to become President Chester A. Arthur's secretary of the interior, Tabor was appointed to fill the remainder of his term. By then, Tabor's roving eye had spotted a young, attractive divorcée, Elizabeth Bonduel McCourt Doe, known to the Leadville miners as "Baby Doe." Tabor and Augusta went through a bitter divorce battle, but Horace won and married "Baby Doe" in an elaborate ceremony in Washington, D.C.'s Willard Hotel, attended by President Chester A. Arthur. Horace and "Baby Doe" lived a lavish lifestyle in their Denver mansion, and she supported the campaign for women's suffrage in Colorado. She bore two daughters, Elizabeth Bonduel Lily Tabor and Rosemary Silver Dollar Echo Honeymaid Tabor.[92]

By the mid-1880s, Tabor's tangled business affairs began to falter. The Matchless Mine was less productive, his investments in real estate and mining were reckless and unprofitable and by 1886, he had mortgaged all his Denver properties. By 1887, he had even mortgaged the Matchless. The Silver Crash of 1893 was the final straw. In March 1894, Horace wrote pathetically to "Baby Doe" from Mexico, where he and her brother Peter McCourt had been developing a mine that proved unprofitable. He reported that Peter was "as badly wrecked as we are but we have our health and the children and will get out all right some way. I have great confidence in you and your ability."[93]

By 1898, Horace, "Baby Doe" and their two daughters were living in a tiny cabin in Ward while Horace tried to develop a new mining property. The *Wray Weekly Times* noted regretfully, "His speculations turned out badly, growing worse year after year until at last he was penniless and an object of public pity." Rescued by Senator Edward Wolcott, Tabor was appointed postmaster of Denver and enjoyed a return of some prosperity before dying of appendicitis in April 1899. His body lay in state at the capitol in Denver, and the viewing drew an estimated ten thousand people. Horace Tabor was silver's greatest beneficiary—and its victim.[94]

As 1893 ended, the fortunes of Colorado's silver industry echoed Horace Tabor's downfall. By September, the Colorado Bureau of Labor Statistics recorded that 377 businesses had failed, 435 mines had closed (about half of all the producing mines listed in 1892) and 45,000 people were unemployed.[95]

Former Kansas senator John J. Ingalls warned darkly of retribution if some relief was not provided during the dire national depression:

> *We are accustomed to speak of this nation as the land of the free and the home of the brave. It will soon be the home of the rich and the land of the slave.*
>
> *A financial system under which more than one-half of the enormous wealth of the country, derived from the bounty of nature and the labor of all, is owned by a little more than 30,000 people, while one million American citizens, able and willing to toil, are homeless tramps starving for bread, requires adjustment. A social system which offers to tender, virtuous and intellectual women the alternative between prostitution and suicide as an escape from beggary, is organized crime, for which some day, unrelenting justice will demand atonement and expiation.*[96]

Chapter 4

DISASTER IN DENVER

By 1893, Denver was a city of more than 100,000 people and had assumed the title "Queen City." Denver began in 1858 at the very beginning of the massive Colorado Gold Rush. Originally two tiny communities, St. Charles and Auraria, Denver City was created when General James Larimer and his backers waltzed in and took over St. Charles, renaming it after the territorial governor of Kansas, James Denver.[97]

With the gold rush frenzy over, Denver secured designation as the territorial capital, and it became the supplier for the burgeoning gold and silver mining industry. This was especially true after rail connections from Denver to the transcontinental Union Pacific line were secured by the Denver Pacific and Kansas Pacific routes in 1870. After that, Denver was the railroad hub and center of commerce for all of Colorado. Construction of ore smelters in Denver accelerated in the 1880s after rail connections to the mining towns meant that ore could be conveniently shipped to Denver and Pueblo.[98]

Wealthier residents built luxurious homes in the Capitol Hill, Montclair and Park Hill neighborhoods. It was in Capitol Hill that J.J. Brown and his wife, Margaret "Maggie" Brown, bought a large, comfortable house in 1894 after J.J. struck it rich at the Little Jonny mine. Working-class citizens lived in far more modest houses, and by 1893, ethnic neighborhoods had crystallized in Denver. The Italians, Poles, Czechs, Germans and Slovenians lived close to the smelters where they worked in the Argo, Swansea and Globeville neighborhoods. One of Denver's "streetcar suburbs," Five Points, attracted many African American residents who worked at nearby railroad yards, and

Margaret and J.J. Brown's home is typical of the fashionable dwellings built in the late 1800s in Denver's Capitol Hill neighborhood. *Library of Congress.*

Fire Station No. 3 in Five Points became Denver's first all–African American fire station in 1893.[99]

Downtown Denver was acquiring buildings of some architectural distinction, beginning with the Tabor Block, Windsor Hotel and Tabor Grand Opera House, all financed by Horace Tabor. Looming over all was the new state capitol building, still under construction in 1893, with its imposing marble flooring, brass fixtures and stained-glass windows.[100]

Through the efforts of Francis Wisebart Jacobs and others, Denver's charitable organizations were well established and strongly supported by the city's elite.[101]

Despite all these admirable efforts to establish a solid basis for Denver charities, the tidal wave of homeless and desperate people flooding into Denver in 1893 overwhelmed the established organizations. Thousands

The Tabor Grand Opera House in Denver, regarded as the most beautiful theater ever built in Colorado, was demolished in the 1960s. *Library of Congress.*

of unemployed miners and their families crowded into Denver when the Panic of 1893 and the Silver Crash hit in full force. In early July, the State Board of Charities held a mass meeting in Denver to organize relief efforts throughout the state. The *West Side Citizen* reported on July 21 that eleven banks and twenty-three major businesses had failed in the previous week alone.[102]

Denver leaders were working with the railroads to offer cheap fares for those who wanted out of the city. By July 28, the *Pantagraph* was reporting:

Colorado's State Capitol was completed in 1895 during the Silver Crash and built completely of Colorado granite, marble and onyx. *Library of Congress.*

> *There are in Denver 10,000 men unable to secure employment, and that there are 20,000 in outside towns in the same condition, one-half of whom are making their way to this city, where they will become a public charge if conditions do not speedily change. In order to avoid this, all the commercial bodies in this city will meet with the representatives of eastern roads this evening and the latter will be asked to make a $5 rate from here to the Missouri river for this class of people. It is argued that these men are not paupers, but if they can get into the agricultural districts they will be able to secure employment in harvest fields, where men will soon be in demand.*[103]

The homeless and unemployed in Denver were so desperate to escape the city that on July 29, 1893, about three hundred unemployed men hijacked a Union Pacific train and headed east. At the state line, a squad of police officers stopped the train. After negotiations, the men were allowed to proceed when they promised not to stay in Kansas City as a drain on that city's resources. When they reached Kansas City, it was reported that they scattered in all directions.[104]

At the same time, several hundred more unemployed men were on their way to Denver from Leadville and Colorado Springs, attracted by the city's free soup kitchens.[105]

In late July, a relief camp was established in River Front Park along the South Platte River. General E.J. Brooks of the state militia was in command of the camp, known as Camp Relief, and it was run along military lines. Homeless men could find lodging in tents, if they had a card signed by Reverend Thomas A. Uzzell of the People's Tabernacle Mission or Dean A.C. Peck of the Haymarket Mission. No liquor was allowed, and no one was permitted outside the camp after 10:00 p.m. A hospital was established, run by the militia's medical officers, and "delicacies" could be donated for the sick. Cooks were recruited from among the homeless men, and five hundred people were fed at each meal twice daily.

In early August, Governor Waite and city officials decided to close Camp Relief. It was reported that "professional tramps" instead of the deserving unemployed were taking advantage of the camp. The city fathers decided to take a different approach, budgeting $15,000 for work relief projects around Denver. Preference for employment was given to married men with families. Only about 170 men out of the 900 who applied were accepted, and the *Avalanche* reported on August 11 that "the 400 men who have recently been fed at Camp Relief, Denver, have been turned upon the streets to rustle and a frightful increase in crime is noticeable at the capital."[106]

With the relief camp closed and the city offering only limited relief work, care of the huge numbers of unemployed in Denver became the responsibility of private charities. The People's Tabernacle and the Haymarket Mission were two of the most active.

People's Tabernacle was founded by one of Colorado's unique and beloved individuals, Reverend Thomas A. Uzzell. A farm boy from Illinois, Uzzell became a Methodist minister and was posted to the raw mining town of Leadville in the early days of the silver boom. When a miner tried to "claim jump" the lot that Uzzell had bought for a church, Uzzell stripped off his coat and offered to fight. The miner ran off, and Uzzell gained a reputation as the

Reverend Thomas Uzzell (1848–1910), the "fighting parson," led the People's Tabernacle, one of Denver's foremost charitable organizations. *History Colorado.*

"fighting parson." Thomas Uzzell's brother Charles founded the People's Tabernacle mission in Denver in 1884. When Charles was forced to resign because of ill health, Thomas took over and spent the rest of his life managing the Tabernacle. The People's Tabernacle offered a bathhouse, free dispensary, free kindergarten (said to be the first in Denver), night school, church savings bank, employment bureau, literary society and alcoholics program similar to today's Alcoholics Anonymous.[107]

The Haymarket Mission was managed by Reverend Arthur C. Peck, usually known as Dean Peck. Born in Ohio, Peck became a Methodist minister and was passionate about education, becoming a trustee and board member of the University of Denver. He solicited enough donations to organize an urban mission in Denver, and it opened in November 1892 in the abandoned Haymarket Theater. The Haymarket Mission offered the same services as the People's Tabernacle, but on a larger scale. By 1894, the mission reported that in the previous year it offered 32,000 meals and 10,000 lodgings, treated 6,000 patients in its dispensary, found employment for 2,632 people and operated the Haven, a facility for homeless girls. Two outgrowths of the Haymarket Mission were the Maverick Restaurant, which seated 150 people and offered five-cent meals three times a day, and the Maverick Woodyard, which gave jobs to 63 men daily, paid in meal and lodging tickets.[108]

The homeless crisis in Denver continued to plague the city throughout 1893. On September 8, the *Pantagraph* reported that dozens of women and children were sleeping outdoors in Lincoln Park each night. The police allowed them to sleep in the park but drove away any men trying to do the same thing. Three men committed suicide in the previous week, and the *Pantagraph* commented, "The condition of the homeless is distressing."[109]

By 1894, a new wave of the unemployed added to Denver's problems with the arrival of the Army of the Commonwealth of Christ, usually called Coxey's Army. The idea of a "petition in boots" came to Jacob Coxey of Massillon, Ohio, in early 1894. A middle-class farmer and quarry owner, Coxey, born in Pennsylvania, had started work in an iron mill as a teenager and later sold scrap iron before moving to Massillon. When the Panic of 1893 hit, he devised a plan along Populist lines, like Franklin D. Roosevelt's 1930s WPA or CCC. Under Coxey's plan, the U.S. government would offer public works employment on roadways and bridges to the unemployed, financed by $500 million in paper currency. This would not only assist the millions of unemployed men but also increase the money supply, create mild inflation and help debtors. The first contingent of Coxey's Army (sometimes known as the Commonwealers or the Industrial Army) started from Massillon on March 25, 1894, with Washington, D.C., as their objective.[110]

Coxeyite groups formed in Denver and many of the mining communities. On April 4, 1894, the *Queen Bee* reported, "A branch of Coxey's Industrial army has been organized in Denver under the direction of Bert Hamilton. They propose to ride to Washington if the railroads can be compelled to carry them." By May 9, the same newspaper stated that the "Colorado Army" consisted of 100 men from Denver and 500 from Cripple Creek. Reverend W.A. Jackson, a former newspaperman and preacher from Pueblo, assumed the position of chaplain for the Denver Coxeyites. The various Coxeyite contingents assembled in Denver and encamped at River Front Park, the former site of Camp Relief. At the urgent request of Reverend Myron Reed and the Denver Coxeyite leaders, Governor Waite agreed to loan tents to be used by the approximately 350 men in the camp by May 24. More men were arriving each day, carrying banners that said "On to Washington," "Cripple Creek Legion of the Coxey Army" and "Give Us the Free and Unlimited Coinage of Silver," among others. Estimates of eventual Coxeyite numbers in Denver ranged from 1,500 to 2,000.[111]

Many Denverites wearied of housing and supplying the Coxeyites, despite the efforts of a volunteer support group called the Home Guard. The Denver Chamber of Commerce arranged for the donation of lumber to the Coxeyites so they could build flatboats and rafts and float down the South Platte River to points east. On June 20, the *Queen Bee* was reporting that no more Coxeyites would be allowed to come to Denver and no further aid would be supplied by the county or the chamber of commerce.[112]

By then, the Coxeyites had decided to leave Denver anyway. In early May, many men started to walk to Cheyenne, where they hoped they could get a lift on one of the railroads. There were ninety-four men walking; eight or ten had been discharged for begging, a practice forbidden by the group's leaders. The *Queen Bee* reported that they were "orderly-looking," under perfect discipline and at least half seemed to be skilled mechanics. At Fort Morgan, each man was given half a loaf of bread, a big chunk of roast beef,

Right: Ohio businessman Jacob Coxey (1854–1951) envisioned a "petition in boots," a march to Washington energizing Congress to provide work for the unemployed. *Library of Congress.*

Below: Unemployed men of Coxey's Army built boats with donated lumber to navigate the Platte River in 1894. *Denver Public Library, Western History Collection X-21556.*

a hard-boiled egg and a fruit can full of coffee. Wagons from Fort Morgan were assembled by the community to take the men to Brush.[113]

Besides the men who walked to Cheyenne, about seven hundred tried to float down the South Platte River in boats and rafts in mid-June. One of the barely manageable boats hit a railroad bridge and capsized. At least six men drowned, but there may have been more casualties. A second attempt was equally futile: only about twenty-seven men managed to finally make it by boat down the South Platte River to the main Platte channel in Nebraska, then to the Missouri River and south to Kansas City. Probably none of them managed to make it to Washington.[114]

Others were still trying to travel by rail. At LaSalle, Colorado, 400 men hijacked a freight train and picked up other Coxeyites but were stopped by a sheriff's posse at the Julesburg city limits. About 500 men camped along the South Platte River near Julesburg waiting for another train to possibly pick them up. Others broke into a railroad roundhouse, stole a locomotive, assembled a train of railroad cars and tried to make it to the east. Railroad workers threw a switch and derailed the train. Then Marshal Jones and 120 deputies arrived, arrested enough men to fill two railroad cars and took them to Denver for trial. The rest of the army split up and tried to catch other trains in Nebraska.[115]

Though the Denver Coxeyites failed to make it past Kansas City, about four hundred weary, dusty, ragged men did arrive in the Washington, D.C. area and encamped in Maryland in late April. When Jacob Coxey and some of his associates tried to present a petition at the Capitol on May 1, they were arrested for trespassing on the Capitol grounds and held for twenty days. Coxey's Army disintegrated by August, and the men tried to make it back to their homes any way they could. The *West Side Citizen* reported on August 31 that several hundred Coxeyites were stranded in Omaha, unable to secure transportation west, and that "the men are scattered throughout the city and are at a loss to know what to do for food."[116]

In Denver, there was outrage at the treatment of Coxey and his men. The *Woman Voter* reported:

> *A set of resolutions denouncing the conviction of Citizen Coxey for the trivial offense of walking on the grass and commending the leader of the Commonweal army for his "humane and praiseworthy mission" was presented and unanimously passed at last Sunday's meeting in Coliseum hall. Over a thousand names were attached to the resolution.*[117]

It was inevitable that some Denverites would turn on one another during this crisis. Many joined the American Protective Association, a fraternal organization dedicated to eliminating the influence of the Roman Catholic Church. Founded in Clinton, Iowa, in 1887, by 1894 the APA boasted of 100,000 members and dominant influence in many midwestern cities where members held political office or controlled those in office. The group bemoaned the fact that most recent immigrants were Catholic and that Catholics—especially the Irish—had inordinate influence in many municipal governments. An APA member named James R. Hoey declared in a letter to the *West Side Citizen*, "Ours is not a war of extermination. It is simply a decisive and unyielding stand against an evil that menaced the world since the first pontiff assumed supreme jurisdiction over all things religious." Hoey also referred to the Catholic clergy as "vipers" and "demons."[118]

Some Coloradans connected the APA crusade against Catholic influence with the crusade against saloons and their evil influence. Claiming that the Catholic Church was allied with the saloons, the *Queen Bee* editorialized in June 1894, "This church and the saloons have ever worked hand in hand. In some sections of the country the contribution boxes of the church are kept tacked up in the saloons. This could be seen at Pueblo, Colorado and at other points. The further south the more brazen."[119]

Conflict between Irish Catholics and APA members in Greeley caused three separate riots. In the last one, an Irishman was shot by one of the APA members. Many APA people carried shotguns and rifles to their meetings. The *Fort Collins Courier* described the APA as "fanaticism run mad."[120]

Despite all the anti-Catholic ferment, Colorado's Irish, German, Italian, Hispanic and Slavic Catholic population was able to withstand the influence of the APA. The APA leadership attempted to obtain support from the Republican presidential candidate in 1896, William McKinley, in exchange for its endorsement. McKinley angrily rebuffed the APA, saying he could not "endorse any abridgement of the constitutional guarantee of religious freedom." By 1898, the APA was fading away, and it was extinct by 1911.[121]

In his history of Denver, Jerome Smiley stated that during the Silver Crash, "opportunities for gaining a livelihood by manual labor and other personal services had sunk to the lowest ebb ever known in Denver. The prairie railroads were forced, through sheer inability to do otherwise, to carry hundreds on their trains without payment of fare.…The exodus from Denver continued in lessening extent through 1894 but may be said to have practically ceased with the close of that year." Smiley estimated that fifteen thousand to twenty-five thousand people left the Denver area during the Silver Crash. He noted that the list of city properties available to be sold because of delinquent taxes filled seventy-three pages of the *Rocky Mountain News*. He referred to 1894 as a year of "demoralization and stagnation" in Denver.[122]

In this time of suffering and crisis, a prophet appeared in Denver—one of the most enigmatic characters in Denver history. His name was Francis Schlatter. Originally from Alsace-Lorraine, Schlatter was a cobbler. He immigrated to the United States in 1884, and by 1892, he had a thriving shoemaking business in Denver. Then he had a vision from "The Father" to sell his business, give the money to the poor and start on a pilgrimage.

Schlatter returned to Denver in September 1895. Tall, bearded and with long, dark hair, he fit the popular image of Jesus perfectly. To some, it seemed that he was the Messiah and that the second coming of Christ was at hand. Tens of thousands of afflicted people stood in line to be healed by Schlatter at his residence in North Denver's Highlands neighborhood. He never asked for any compensation for his healing efforts.[123]

On November 13, Schlatter left a note saying that his mission had ended and that he was leaving Denver. He stayed at the Morley Ranch in New Mexico for several months and then left, saying that God was establishing a new Jerusalem in the New Mexico mountains. In 1897, a skeleton, presumed to be Schlatter's, was found in the Datil Mountains, and some of his effects were discovered in a nearby village. It was assumed that he perished alone in the mountains, but no definitive answer has ever been learned about the fate of the man who many hailed as the new Messiah.[124]

In 1895, the Colorado economy began to revive with the new gold discoveries in Cripple Creek and Victor. Even silver mining saw an increase in production. To signal this to the nation, Denver civic leaders latched onto a suggestion from Denver and Rio Grande Railroad president Shadrick K. Hooper, known as "Major" Hooper. Hooper's recommendation to Denver was to hold a massive community celebration like Mardi Gras in New Orleans, celebrating Colorado's history and what the state created and produced.[125]

Denver civic leaders immediately responded by planning a massive citywide event called the Festival of Mountain and Plain. All Colorado cities and industries were invited to participate. Four massive themed parades with more than one hundred floats wound their way through Denver, and other celebrations included Native American encampments and dances, Wild West shows, masquerade balls, band contests and single-handed and team rock drilling. The festival was a huge success, with an estimated 150,000 visitors from all over Colorado and the rest of the country. The festival was held again through 1912. The Festival of Mountain and Plain had the

Right: Many saw Francis Schlatter (1856–1896?) as the Messiah with supernatural healing powers, especially since he fit the popular conception of Jesus' appearance. *Library of Congress*.

Below: A Denver crowd waits patiently in the rain in 1895 for the opportunity to be treated by famed faith healer Francis Schlatter. *Library of Congress*.

This extravagant Zang's Brewery float and others like it amazed and delighted Denver residents during the Festival of Mountain and Plain from 1895 to 1912. *History Colorado.*

distinction of being the subject of the first movies ever filmed in Colorado, by the Edison Motion Picture Company in 1897. This film footage is in the collection of the Library of Congress. The name of the festival is still in use as one of the alternate names of the annual Taste of Colorado event held in Denver's Civic Center Park.[126]

The Festival of Mountain and Plain was a grand celebration, but it only camouflaged the fact that Colorado was still struggling to completely recover from the Panic of 1893 and the Silver Crash. On June 1, 1895, the *Boulder Daily Camera* reported, in a story headlined "Homeless in Denver," that two hundred people were forced out of Jerome Park in Denver by flooding from Cherry Creek. The festival also glossed over political and economic divisions that were to almost bring civil war to Denver and Colorado and doom the stormy political career of Colorado's Silver Crash governor, Davis H. Waite.[127]

Chapter 5
BLOODY BRIDLES

When Davis H. Waite was inaugurated as governor in January 1893, he rode to victory on a tide of enthusiasm for the Populist Party. Colorado voters elected thirty-nine members of the Populist Party to the state legislature, as well as giving most of their presidential votes to James Weaver, the Populist candidate. When Weaver and Mary Lease spoke in Denver in August 1892, an audience of ten thousand people assembled, said to be the largest Populist gathering in the country except for the national convention.[128]

But soon after assuming the governorship, Waite was forced to deal with the dire consequences of the Panic of 1893 and the Silver Crash. Davis Hanson Waite was sixty-seven years old in 1893 with white hair, a long white beard and gold-rimmed spectacles. Despite his patriarchal appearance, he was quick to anger and often acted intemperately without consulting with others. Possibly the best evaluation of Waite was printed after his death by the Durango *Wage Earner*, describing Waite as "a kind-hearted and honest man, but of a strange, erratic temperament."[129]

Despite these drawbacks, Waite earned political praise by supporting the drive for women's suffrage. In November 1893, women earned the right to vote in the state of Colorado in all elections, including national ones. There had been efforts to secure women the right to vote as far back as 1870. By that year, Utah and Wyoming Territories had already authorized female suffrage through legislative action. But in Colorado, the campaign for women's suffrage was won through popular referendum, not by a state or territorial legislature. Previous efforts to gain women the right to vote in Colorado failed in 1876 and 1877.[130]

By 1893, the political landscape in Colorado had shifted. Representative J.T. Heath, a Populist from Montrose and Delta Counties, introduced the proposal for a women's suffrage referendum into the Colorado legislature in January. His bill was approved by the Colorado House and Senate and placed on the ballot. The result was a smashing victory for women's suffrage at the ballot box, making Colorado the first state in which a vote by the people freed women to express their opinion at the polls. This was also a victory for Governor Waite, since he and the Populists had allied themselves so strongly with the women's suffrage movement.[131]

Governor Davis H. Waite (1825–1901) tried alleviating the Silver Crash with controversial solutions and protected miners during the Cripple Creek strike of 1894. *History Colorado.*

Waite was faced with a crucial issue with the influx of thousands of unemployed men and their families into Denver after the Silver Crash began in the summer of 1893. His short-lived attempt to help with the establishment of Camp Relief proved too expensive, and it attracted too many "professional tramps" instead of the deserving unemployed. Certainly, his drive to build the new state capitol building with Colorado granite and marble helped provide some additional employment. But he searched for something that would offer much more statewide relief.[132]

Frustrated by the lack of progress, Waite gave a fiery address to an overflow audience at a Silver Convention in Denver on July 11, 1893, an address that would brand him with the nickname he would never shake. He thundered out:

> *Who is Grover Cleveland and who is Benjamin Harrison, and who are their supporters on Wall Street or in Denver that in this nineteenth century they dare assume to drive into poverty and exile a half million of American freemen?...And if the money power shall attempt to sustain its usurpation by the "strong hand" we will meet that issue when it is forced upon us, for it is better, infinitely better, that blood should flow to the horses' bridles, rather than our national liberties should be destroyed.*

Women wait to vote after an 1893 referendum granted women the right of suffrage in all Colorado and national elections. *Wikipedia*.

Ever after, Davis Waite would be known and ridiculed as "Bloody Bridles" Waite.[133]

By late in 1893, Waite hit upon the idea of calling the Colorado legislature into special session to approve a plan for using scrip to enhance the state's money supply. Scrip was used by mining companies and other corporations and on a limited basis by cities and counties. It was basically a form of money certificate that could be used to buy goods at company stores or be redeemed in cash by cities and counties in exchange for work performed. Waite's plan was described by the *Aspen Weekly Times* in January 1894. The legislature would be called into special session to provide work for the unemployed and to "permit state, county, cities and school districts to make public improvements on their own credit, with their own scrip, and that scrip receivable for taxes." To some, this seemed like a credible solution to the Silver Crash, in effect allowing for a more flexible flow of money. In July 1894, the *Queen Bee* editors wrote in favor of the scheme, saying, "Let Colorado make her own money. If corporations can issue scrip payable for groceries, why cannot the state do the same thing."[134]

Waite cast about for ideas about implementing his scrip scheme, writing to Mayor William Cowherd of Kansas City about that city's experience at issuing scrip. The answer was discouraging; though the redeemed, used scrip was supposed to be burned, someone made off with a big bundle of it, and the city ended up paying it back twice to the tune of about an extra $100,000. The chaos that would be caused by having all kinds of organizations issuing their own scrip of different values was also pointed out.[135]

Instead of allowing wholesale issuing of scrip, Waite hit on another plan: Colorado would send its silver bullion to Mexico to be minted into silver Mexican dollars, which would then be returned to Colorado and used as legal tender. This would offer a use for Colorado's surplus silver, open the mines again and provide a plentiful supply of silver currency to expand the state's money supply. The proposed hybrid Colorado/Mexican silver dollars became known derisively in Colorado as "fandango dollars." The *Idaho Springs News* explained Waite's plan:

> *The old coinage laws by which foreign silver coins are legal tender are understood by the governor never to have been repealed and he believes that Colorado silver coined in the Mexican mints can legally pass current here. The governor is even said to have sent a personal representative to the City of Mexico to confer with the treasury officials there and complete the details of the project.*[136]

Critics immediately jumped on the fandango dollars scheme as illogical, impractical and illegal. In an address to the Manufacturers' Exchange, ex-governor James B. Grant raked Waite over some very hot coals: "Under the circumstances it is surprising that outside of an insane asylum a man can be found with so many wheels in his cranium as to even suggest a state solution of a question so entirely and so absolutely national in its scope and character as is the coinage of gold or silver."[137]

All the Denver newspapers thunderously condemned the fandango dollars scheme. The *Rocky Mountain News* did it in colorful terms: "Colorado's governor insists that the state legislature shall make a monkey of the state, monkeys of its people, and double-tailed monkeys, ourang outangs and gorillas combined of the party that elected him to the office he holds." In more dignified terms, the *Denver Republican* simply called Waite's scheme "a gross travesty on common sense."[138]

When a United States Treasury official was asked about the scheme by an Associated Press reporter, he replied, "What would the profit on the transaction amount to after the transportation charges had been paid on the bullion to Mexico and the coins on their return to the United States?"[139]

Governor Waite did call the legislature into special session on January 10, 1894. After several days of vigorous debate, the fandango dollars scheme was overwhelmingly rejected on January 12 by a vote of forty-four to eighteen in the State Senate. Smarting from the defeat of his pet scheme, within a few months Waite would react to a crisis in Denver in a way that would bring him even more condemnation.[140]

Denver's underworld in 1894 was controlled by three rival gangs, sometimes partnering, more often jockeying for the best position. The oldest and best-established was run by "Big Ed" Chase, who had been in Colorado ever since the gold rush days. "Big Ed's" gambling and prostitution empire included high-end houses of soiled doves such as the Navarre, gambling palaces like the Progressive and burlesque theaters like the Palace and Mascot. Jefferson Randolph "Soapy" Smith (so called because of his fake raffle of soap packages, one of which supposedly contained a $100 bill) arrived in 1879 and organized his own powerful combination of con men, swindlers and musclemen. The Blonger brothers, Lou and Sam, slimed onto the Denver scene in 1889 and showed themselves masters of the elaborate "big con," targeting tourists with fake stock exchanges, betting offices and auction houses. Many city officials, police officers and even one congressman, Robert Bonynge, were paid off by one or more of the gangs.[141]

Denver's corrupt political machine was influenced not only by criminal gangs but also by powerful financial interests and individuals controlling all the city's utilities and transportation. Reformer judge Benjamin Barr Lindsey referred to this whole combination as "The Beast." Walter Cheesman, William Gray Evans, E.B. Field, David Sullivan and David Moffat headed up the Denver water, tramway, gas and electrical and telephone companies, railroads and banking interests. Through distribution of plum city jobs, bribery and intimidation, they controlled Denver's city government.[142]

A peculiarity of the new 1893 Denver city charter was that the governor of Colorado could appoint a certain number of officials to the Denver Fire and Police Board. On March 8, 1894, Governor Waite removed two of the board members, Jackson Orr and David J. Martin, for allowing illegal gambling to proceed without interference. He replaced them with Dennis Mullins, a Populist, and Samuel D. Barnes, a Republican. This ignited what became known as the "City Hall War." Orr and Martin applied for, and received, a ten-day injunction from Denver District Court restraining Governor Waite, Mayor DeKalb Van Horn and the two newly appointed board members from interfering with their jobs. They attempted to have the injunction made permanent, at the same time barricading themselves in their offices. The entire Denver Police Department was called in to defend city hall. Several hundred "special deputies" were also sworn in and armed with two revolvers each. Some of these men were described as "man killers," and legend says "Soapy" Smith was among them.[143]

The whole situation exploded on March 15, when Governor Waite lost patience, declared Denver to be under martial law and called out the militia

Left: "Big Ed" Chase (1838–1921), Denver's vice lord, controlled a gambling, prostitution and con game empire and wielded tremendous influence over city politics. *History Colorado*.

Below: Jefferson Randolph "Soapy" Smith (1860–1898), Denver's master con man and grifter, in a saloon in Skagway, Alaska, 1898. *Library of Congress*.

Colorado State Militia troops await orders during the 1894 City Hall War, with two Napoleon cannons behind them. *Denver Public Library, Western History Collection X-22120.*

to attack city hall. He also made a request for federal troops from Fort Logan. A militia infantry company, a Signal Corps company armed as infantry and the Chaffee Light Artillery with two Gatling guns and two outdated muzzle-loading Napoleon cannons formed in a battle line in front of city hall. The police and deputies, by now all armed with rifles and shotguns, also prepared for battle on the main floor of city hall. Even the Denver Fire Department joined in, positioning fire hoses to rake the militia troops if they tried to attack. A crowd of thousands gathered to see what would happen if open warfare erupted.[144]

A deputation of prominent citizens including chamber of commerce officials negotiated with the old board members and Governor Waite, but it seemed that neither side would yield. Federal troops of the Seventh U.S. Infantry Regiment arrived and assembled at Union Station. Numerous deadlines were given to the old board members to vacate city hall, but each one was ignored. Governor Waite received death threats and was put under guard. Colorado Militia adjutant general Thomas J. Tarsney, a Populist, said if he were ordered, he would blow city hall to pieces. Seven ambulances were drawn up, ready to convey the wounded to hospitals. It seemed that at any moment a massacre would bathe the Denver streets in blood.[145]

After a conference with Adjutant General Tarsney, Governor Waite finally ordered the militiamen back to their armory at 9:00 p.m. The old and new

A crowd of thousands flocked downtown during the 1894 City Hall War in ghoulish anticipation of seeing mayhem. *Denver Public Library, Western History Collection X-22118.*

board members agreed to submit the situation to the courts for a decision. The police, firemen and deputies stood down. On March 16, the Colorado Supreme Court decided that Governor Waite did have the right to replace the old board members, but not for political reasons, and that he did not have the authority to install the new board members before the court injunction against it had been adjudicated. He was also condemned for attempting to use the state militia to enforce his decision. On March 26, Judge James Glynn of Denver District Court voided the earlier injunction and stated that Orr and Martin were guilty of censurable action in fortifying city hall. This was the final decision, a decidedly mixed bag for Governor Waite. *The Avalanche* sneered that the Supreme Court justices must have "giggled audibly" in laying down a decision that damned Waite with faint praise.[146]

The threat of violence in Denver was quickly overshadowed by actual violence in Cripple Creek. On March 17, Governor Waite received an urgent telegram from Sheriff M.F. Bowers of El Paso County, saying that two of his deputies had been shot that night, that a mob of five hundred men had gathered at Altman and that state militia troops were needed to restore order. Waite immediately ordered the Denver militia companies to report to their armory and the Chaffee Light Artillery company to march to Union Station for embarkation.[147]

In the aftermath of a bitter mining strike in Coeur d'Alene, Idaho, representatives from Idaho, Montana, Colorado, South Dakota and Utah met in Butte, Montana, and on May 15, 1893, founded the Western Federation of Miners of America. Cripple Creek–area miners formed Free Coinage Miner Union No. 19 of the Western Federation of Miners to protect their interests, electing John Calderwood president of the union.[148]

By January 1894, a labor dispute was brewing in the Cripple Creek area. The *Aspen Weekly Times* reported that the mine owners were demanding a move by February 1 from an eight-hour working day to a nine-hour day with no increase in pay, at three dollars per day. On February 2, 250 miners marched to the Victor mine and prevented other miners from entering. On February 10, the primary mine owners met in the Summit Company offices in Colorado Springs and decided not to yield on a nine-hour, three-dollar day and to keep the mines closed until the miners agreed. So, the situation in Cripple Creek became part strike and part lockdown. On February 2, David Moffat was quoted in the *Herald Democrat* as saying, "I will not be dictated to, and if the men are not satisfied with my terms I will close down every mine I have in the camp, and keep them closed until I can run them peacefully, and on my own terms, even if it takes me five years to do so."[149]

From this point on, the Cripple Creek strike built to the level of a civil war in Colorado. The mine owners obtained an injunction from the district court in Colorado Springs to prevent the striking miners from interfering with the operation of the mines on March 14. When six deputies were sent by Sheriff Frank Bowers to enforce the injunction on March 16, they were fired on by miners; one was wounded, and they were all captured. The deputies were taken to Altman, hauled before a judge, disarmed and released. It was then that Sheriff Bowers sent his telegram to Governor Waite calling for intervention by the state militia.[150]

Waite rushed several militia companies, about three hundred men, to the Cripple Creek area. The militia commanders found the situation quiet if tense and recommended that the militia be withdrawn. Waite agreed, and on March 20, the militia returned to Denver. On March 31, the *Aspen Weekly Times* reported that the condition of the working men in Cripple Creek was "deplorable." There were two thousand or more idle men, almost all businesses except saloons and hotels had shut down and "hundreds of men are without money or shelter other than saloons where they are allowed to bunk down at night."[151]

A conference was held in Colorado Springs in early May between the mine owners and the union. The owners offered to institute an eight-hour workday at $2.75. The union refused the offer. At this point, the owners decided to begin hiring strikebreakers to replace the striking miners. They also began hiring additional deputies to protect the strikebreakers against harassment or violence, with Sheriff Bowers in command. The *Fairplay Flume* reported on May 31 that 105 men, many of them former policemen or firemen who had been discharged by the new Fire and Police Board, left Denver for Cripple Creek, where they would be deputized by Sheriff Bowers. An additional 125 men were recruited from Denver, Colorado Springs, Pueblo and Leadville and would soon join them.[152]

In response, the striking miners organized along military lines. It was reported that two West Point graduates were among the miners. One of them, Junius Johnson, organized the miners into companies, drilled them in military tactics and picket duties and created a fortified camp on Bull Hill overlooking Altman. Strikers circulated around all the mining communities in the area, confiscating firearms and leaving receipts so the owners could retrieve the firearms after the conflict ended. The *Aspen Weekly Times* noted, "The miners are wrought up to a pitch that the least provocation would be like touching a match to tinder." When union president John Calderwood was absent at a mining convention in Salt Lake City, violence erupted.[153]

As the deputies arrived from Denver on the Colorado Midland Railroad on May 25, the striking miners blew up the nearby Strong Mine with dynamite, trapping three men inside and showering the deputies with debris. Many of the miners mingled with recently discharged railroad workers who were blowing their paychecks in local saloons, and a wild, chaotic, drunken spree ensued. The deputies prudently withdrew a few miles away, but a trainload of miners followed and sent scouts out to find the deputies and attack them. The deputies ambushed the advancing miners, a gunfight broke out and one man was killed on each side. Junius Johnson managed to get the situation under control and drag the miners back to their Bull Hill camp.[154]

When John Calderwood returned, he arranged to close the saloons and restore a vestige of peace and quiet. A group of 50 volunteer miners was recruited to keep order and expel anyone who got out of control. Fifty Cripple Creek businessmen were arrested by the miners under suspicion of having aided and helped finance the deputies. But volunteer deputies were arriving daily to add to Sheriff Bowers's force, and he eventually assembled a private army of 900 to 1,200 men armed with Winchester rifles and an outmoded Civil War Parrott cannon.[155] Attempting to defuse the situation,

Mine owners hired these African American deputies during the 1894 Cripple Creek strike to attack striking miners. *Denver Public Library, Western History Collection P-1929.*

During the 1894 Cripple Creek strike, striking miners blew up the Strong Mine, temporarily trapping three employees. *Denver Public Library, Western History Collection X-62571.*

Governor Waite traveled to Altman and met with the miners at Bull Hill, where they agreed to allow him to represent them in negotiations with the mine owners. On June 2, he met in Colorado Springs with John Calderwood of the miners' union, J.J. Hagerman of the mine owners and several others. Learning that Waite and Calderwood were in town, a lynch mob assembled and surrounded the building. Judge Horace Gray Lunt addressed the crowd, trying to calm them, and this gave Waite and Calderwood time to escape through another door and make their way to the train station. Otherwise, they very probably would have been lynched or shot.[156]

Two days later, Hagerman and David Moffat went to Denver and met with Waite. They agreed to a compromise settlement of the strike. The mine owners agreed to an eight-hour workday at three dollars pay per day, with a twenty-minute lunch break. There would be no discrimination in hiring union or nonunion workers. The next day, Waite ordered the entire force of Colorado state militia to Cripple Creek to keep order and prevent violence. The eight hundred troops were under the command of General Brooks and Adjutant General Tarsney. Waite issued a proclamation ordering both the miners and the deputies to disarm and disband.[157]

On June 5, the miners at Bull Hill laid down their arms, called in their pickets and prepared for a peaceful resolution of the strike. A huge, flag-bedecked celebration began in Cripple Creek. But then word arrived that Sheriff Bowers was advancing with his deputies, holding two hundred arrest

Striking Cripple Creek miners built "Fort Dean" atop Bull Hill to resist armed intervention by the mine owners' hired deputies. *Cripple Creek District Museum.*

State militia troops called to Cripple Creek during the strike set up "Camp McIntyre," named after Governor Albert McIntyre. *Cripple Creek District Museum.*

warrants against various miners, and that he vowed to make the arrests "at all hazards." The whistle at the Victor Mine was blown as an alarm, and the miners hurried back to their Bull Hill defensive lines and prepared for an attack. Each man was armed with a rifle or shotgun, and the approach to the fort had been mined with dynamite that could be ignited via electric detonating wire. The miners also had a homemade "bow gun" that could hurl beer bottles filled with dynamite down the hill at the deputies. About eight hundred miners were reported to be at the Bull Hill fort.[158]

When it became clear that the deputies were advancing to attack the miners' camp, General Brooks ordered the troops to form in between the two forces. Sheriff Bowers told General Brooks that he was determined to enforce the arrest warrants. Brooks replied that his troops would fire on the deputies if they tried to attack the miners. At this point, Bowers ordered the deputies to retreat.[159]

The enraged deputies marched through Victor, arresting fourteen miners who shouted insults at them. They encamped near the Independence Mine and refused to disband. However, the following day, a conference was held with Sheriff Bowers, mine owners and militia officers. An agreement was reached by which the deputies would disband, the mines would return to control by the owners, the military would protect the mines and miners, the militia would remain for thirty days and the miners would give up any personal property in their possession. The deputies were disarmed and taken into custody by General Brooks. This seemed to offer a final solution defusing the threat of violence.[160]

But that was only an illusion.

In late June, Adjutant General Tarsney was in Colorado Springs, attending the testimony concerning the miners who had surrendered to Sheriff Bowers, had been arrested and had been brought to Colorado Springs for trial. Tarsney was acting as an attorney for the miners, a questionable role for a high-ranking militia officer. On June 22, Tarsney was abducted from his hotel by a group of masked men. In a shocking episode worthy of the antebellum South, he was taken three miles out of town, beaten, stripped and tarred and feathered. He managed to make his way to a ranch house and then into town, then boarded a train for Denver. Governor Waite offered a $1,000 reward for the capture of the offenders. Most were eventually arrested and put on trial.[161]

In the aftermath of the City Hall War and the Cripple Creek strike, Waite was the target of almost universal condemnation. Writing of the strike, the *Colorado Daily Chieftain* editors said, "The question now naturally arises, Who is responsible for this unfortunate condition of affairs? No one but Davis H. Waite, governor of Colorado. To the knavishness and imbecility of this man is due nine-tenths of the trouble....During each visit he made to Cripple Creek he...in some instances openly endorsed the lawless acts

The rear guard of the deputies' army prepares to leave at Midland after settlement of the 1894 Cripple Creek strike. *Cripple Creek District Museum.*

The state militia in front of the National Hotel guard Cripple Creek miners arrested after the strike. *Denver Public Library, Western History Collection X-60252.*

of those engaged in the strike." The Creede *Candle* denounced the Waite administration in colorful terms, saying, "Paramount to all other questions it is necessary for the people to get rid of this life-sucking, disorganizing, infamous horde of political deformers who have been wrecking the state's standing and credit, and connecting the silver cause with such fanaticism in eastern eyes for two years past." The *Meeker Herald* condemned Waite as a troublemaker, saying, "For the past two years he has been engaged chiefly in raising hell and has been more successful in that occupation than in anything else he has undertaken during the seventy years of his life." A personal dig at Waite's often bombastic oratory came from the *Colorado Weekly Chieftain*: "Turpin's new war machine will shoot 25,000 bullets per minute. We now have a better shooter than that up at Denver. It is the governor's mouth."[162]

Waite had also expressed sympathy for the railroad workers involved in the national Pullman Sleeping Car strike led by Eugene Debs. This strike ignited violence across the country, halted much rail traffic, led to the insertion of federal troops by President Cleveland and landed Debs in prison, not for the last time.[163]

Added humiliation was heaped on Governor Waite when he was arrested in the late summer of 1894 for intercepting and retaining a letter to Sadie

Colorado State Militia troops pose with their fearsome Gatling guns at Camp McIntyre during the 1894 Cripple Creek strike. *Cripple Creek District Museum.*

Likens, who was a matron at police headquarters in Denver. The confusing case involved a salacious invitation to Likens from a Pueblo man. The case, the Leadville *Herald Democrat* noted, "is only for political purposes and will amount to nothing." The prediction proved true when the case against Waite was dismissed.[164]

Winfield Scott Stratton (1848–1902) compromised with the union during the 1894 strike, reopening his Independence Mine at three dollars per day. *Cripple Creek District Museum.*

Although Waite was again nominated for governor, he was soundly defeated in November by Republican Albert McIntyre with a majority of nearly twenty thousand votes.[165] Waite's governorship, with the fandango dollars fiasco, the City Hall War and the violent Cripple Creek strike, left an impression of radical anarchy at worst and an unbridled temper and incompetence at best. It was also true that powerful economic interests were antagonized by Waite's Populist rhetoric and his actions defending the miners during the Cripple Creek strike. Nevertheless, Waite was lionized by later writers sympathetic to the plight of labor in the Gilded Age. Emma Florence Langsdon wrote in 1900: "Governor Waite occupies the unique position in history of being the only governor in the United States who ever ordered out state troops to protect the rights of the working man against his oppressor."[166]

Chapter 6

THE MAELSTROM

The fifty years between 1880 and 1930 saw violent clashes between labor and management on a national scale. Workers increasingly unionized to resist demands by owners, and strikes became more violent as workers clashed with strikebreakers and company guards hired to protect them. Federal troops and state militias were routinely called out to suppress violence, usually allied with the corporations. The result was the radicalization of many union members, formation of the Industrial Workers of the World (IWW) and even more violence culminating in the Colorado Labor Wars of 1903–4 and the 1914 Ludlow Massacre.

The Silver Crash destroyed normalcy in Leadville. The sudden drop in silver prices threw thousands out of work, slashed income for mines and smelters and caused dozens of businesses to close their doors. But in November 1893, Leadville started to climb out of the worst of the Silver Crash when huge deposits of gold were discovered in the Little Jonny mine. The resulting resurgence in gold mining restored some measure of prosperity to the Leadville area. The mines gradually began to open again, but because of the still-drastic economic situation and low silver prices, many mine owners cut wages from $3.00 a day to $2.50.[167]

By this time, the union situation in Leadville had changed. In 1892, violence erupted during a vicious strike in the gold, silver and lead mines of Coeur d'Alene, Idaho. The furious reaction was the 1893 formation of the Western Federation of Miners (WFM) in America. Soon, the WFM had two hundred locals spread across thirteen states. Leadville miners joined the new union, forming Cloud City Miners' Union No. 33.[168]

Leadville's toxic industrial environment, sustained by intensive physical labor, is clearly illustrated by this image of the Leadville lead smelting plant. *Library of Congress.*

The WFM initiated the massive 1894 strike in Cripple Creek, resulting in several deaths and much destruction of mining property. Conservative Coloradans saw the union as a radical and dangerous element. The backlash against the violent strike drove Governor Davis H. Waite out of office, along with many other members of the Populist Party. The WFM did take an aggressive attitude toward mine owners. Mine ownership was also changing. Ownership was concentrating in fewer hands, and there were more absentee, corporate owners who had not started out as prospectors and who had little contact with the men who worked the mines. The perfect situation for another long, bitter strike was forming. Many Leadville mine owners had restored the previous daily wage of $3.00, but about one-third of the miners were making $2.50.[169]

In late May 1896, the miners' union met with some of the mine managers, asking that all miners be returned to a daily wage of three dollars. The managers rebuffed the union representatives at this meeting and another

David Moffat (1839–1911), a leader in Colorado railroads and mining, was one of the most stubborn mine owners during the 1896–97 Leadville strike. *History Colorado.*

one held on June 19. That evening, about 1,200 union members met and agreed that all members who made less than three dollars a day would go on strike that night. The mine owners and managers then locked down the Leadville mines, making the Leadville situation, as in Cripple Creek in 1894, a combination of strike and lockout.[170]

Reacting quickly to the strike, the Leadville mine owners met in secret and formed the written pact that would eventually defeat the union. They agreed that no concessions would be adopted without a majority vote of the owners and that none of them would negotiate with the union or even recognize its existence, instead pretending that they were negotiating with individual miners. This was in stark contrast to the situation in Cripple Creek, where individual mine owners such as Winfield Scott Stratton met with the union and made separate agreements to open their mines. The leaders of the miners' union were also intransigent and refused to compromise, perhaps buoyed by the WFM victory in the Cripple Creek strike and the fact that the county sheriff, M.H. Newman, and his deputies were sympathetic to the union. They were also supported by the national WFM leadership in Butte.[171]

The union leaders might have been apprehensive had they assessed the situation more objectively. Most Leadville officials, elected on the Citizens' Ticket, were allied with the mine owners, as were the city police and business leaders. Colorado's governor, Albert McIntyre, was a Republican and not as sympathetic to labor as Populist "Bloody Bridles" Waite (though not openly anti-labor). The mine owners and managers were quite prosperous even during the Silver Crash, and such people as David Moffat, "Leadville Johnny" Campion and J.J. Brown had millions at their disposal to withstand the strike. Another factor was the lack of national support. The American Federation of Labor refused to support the Leadville strike, claiming that its leadership had not been consulted before the strike was called. A separate AF of L union was formed in Leadville during the strike as Cloud City Stationary Engineers, Firemen, Pumpmen and Mine

Harrison Street was the center of Leadville's thriving downtown commercial district in 1896, and most of its imposing buildings still stand. *History Colorado.*

Mechanics Union No. 1. This group had no affiliation with the miners' union and was under no obligation to cooperate with it.[172]

"Leadville Johnny" Campion of the Ibex Mining Company hired spies from the Pinkerton and Thiel detective agencies to infiltrate the union, posing as miners. They were extremely successful, providing the mine owners with detailed reports on all the union plans and activities. Campion had the right to be apprehensive about the union, since it was known that in July it received a shipment of one hundred deadly Marlin lever-action rifles. Five of them were given to Sheriff Newman's union-friendly deputies, and the rest were distributed among union members. Several members of the union's "Committee of Twenty" coordinating strike activities opposed buying the rifles but were ignored.[173]

In mid-August, the mine owners and managers made an offer to the striking miners, as reported by the *Aspen Tribune*: "For all calendar months in which the average silver quotation shall be 75 cents per ounce, the miners shall be paid $3 per day. We offer to make this modification in spite of the fact that silver is likely to advance to 75 cents within a very short time, confident that this offer will appeal to the individual intelligence of the miners." In fact, it did not appeal to the miners in any way, since the pay rate could change at any time, tied to the price of silver. At a crowded meeting on August 19, the union angrily rejected the offer, referring to the group of owners and managers as the "Managers' Union." They stated, "Be it resolved that we, the miners of Leadville, hereby enter our protest against the conspiracy now being carried out by the 'Managers' Union' as contrary to the spirit of our constitution, a violation of our rights as American citizens, a restriction on the individual liberty of action of miners and of managers, immoral, illegal, inhuman and unChristianlike."[174]

The mine owners responded, "If the miners do not make application for work on or before August 22, the companies represented will consider their proposition rejected and no longer binding upon them in any particular." The owners shut down the pumps in two of the leading mines, the Bon Air and Penrose, which would lead to flooding. The *Colorado Daily Chieftain* commented, "This means that the mine owners hope to force the union to declare the strike off, or they think to avoid trouble by closing down on everything and allowing the mines to fill with water, rather than submit to the decision of the union." At the Coronado Mine, a wooden stockade was built around the mine buildings, and wagonloads of provisions were delivered to the compound, as if to withstand a siege. It was obvious that the Coronado was going to open for work again.[175]

Miners like these men from San Miguel County felt they had no choice but to unionize to resist mine owners' attempts to cut wages. *Cripple Creek District Museum.*

With neither the owners nor the union willing to compromise, bitterness and hatred enveloped Leadville. The union formed groups of "regulators," armed with revolvers, who would circulate around Leadville threatening non-union workers, accosting and beating them with revolver butts. The violence grew after the Coronado and several other mines opened on August 26 at a $2.50-a-day pay rate. R.B. Estey, manager of the Coronado, was followed by a rock-throwing mob of women and children. Several men who worked at the Coronado were savagely beaten, and at least one, Frank Hassett, was called a "son of a bitch of a scab," shot and wounded in the leg. The men working at the opened mines were usually local Leadville men and not outside strikebreakers. At the Coronado, they were armed with two Winchester rifles and six shotguns. Judge Owens was alarmed at the threat of violence and predicted that the whole town might be burned down. He told Sheriff Newman of his fears, and Newman sent ten special deputies to the Coronado to guard against trouble.[176]

The top law enforcement officer in Lake County, Michael Newman, was encountering legal problems of his own. Newman operated a barbershop in Leadville prior to his election and was voted into office as a Populist candidate. Less than two months into his term as sheriff, he was arrested in March 1896 as the result of a sting operation. He was shaking down the gambling establishments in Leadville, asking up to $500 a month to ignore gambling regulations and keep them open. Released on bond, he was again arrested in April on the same charge but through a different complainant. A change of venue was granted, and he went on trial in Park County at Fairplay in June on charges of blackmail, oppression and extortion. Newman was convicted and sentenced to removal from office and six months in the Lake County jail. The sentence was appealed and eventually went all the way up to the United States Supreme Court.[177]

It was exactly while Sheriff Newman was dealing with these criminal charges that he was also supposedly managing law enforcement during the strike in Leadville. The fact that Newman was a Populist led to charges of favoritism on behalf of the strikers. Colorado's governor, Albert McIntyre, communicated with Newman in July, saying that he had heard workmen had been stopped in the performance of their jobs, men were forcibly prevented from coming into Leadville and committees were meeting all incoming trains and searching them for strikebreakers. McIntyre offered to activate the state militia to maintain order. The *Weekly Star-Times* reported that "in reply Sheriff Newman said there was no trouble here at all, and that in case there was he was perfectly able to handle any difficulty himself." Newman was tragically mistaken, biased or negligent in his reading of the Leadville situation.[178]

By mid-September, everyone in Leadville was anticipating some sort of violent incident. The union leadership issued a statement on September 17 saying that anyone breaking the law during the strike was "committing treason to our cause." The Coronado Mine was within Leadville city limits, only about six blocks from the post office, and surrounded by houses. At 1:00 a.m. on September 21, a fusillade of gunfire rang out and bullets started hitting the stockade surrounding the mine. The workers inside returned fire. After a twenty-five-minute gun battle, the attackers started throwing lit sticks of dynamite inside the mine compound. Several fuel oil storage tanks east of the engine house exploded and burst into flame, engulfing much of the mine complex in sheets of fire. A fire truck raced up, but the firemen were attacked by the union men, and a heroic fireman, Jerry O'Keefe, was shot and mortally wounded while trying to attach a fire hose to a hydrant. Sheriff Newman tried to singlehandedly stop the fighting but was stopped by a group of miners who threatened to shoot him if he persisted. With the Coronado buildings a blazing inferno, the workers inside were able to escape through the maze of machinery and equipment.[179]

Leadville's Coronado Mine was attacked by striking miners in September 1896, causing the deaths of six men and massive damage. *Lake County Public Library.*

Leadville's militia companies raced to their armories, grabbed weapons and assembled to march to the Coronado. The streets were full of rushing men and crying women and children. Hundreds of men rushed to the mine with firearms, guarding the firemen and allowing them to finally start working. By that time, though, the mine buildings were smoldering ruins and four houses around the mine had burned. All the mine whistles and several steam fire sirens in town were blowing. With the cacophony of shouting men, screaming and crying women and children, shots ringing out, whistles and sirens blowing and the fires raging, Leadville was a scene from hell that night.[180]

At about 3:30 a.m., shots were heard from the Emmet and Marian Mines, located about a mile and a half outside town. Both mines had recently opened at a daily pay rate of $2.50. More than one hundred shots were heard from the second attack. Two other mines, the El Paso and R.A.M., were also attacked unsuccessfully. The mine defenders were able to hold out without any casualties. Sheriff Newman telegraphed Governor McIntyre, saying, "The situation here is beyond my control. I am compelled to call on you for aid to preserve order and hereby do so. Am of present opinion strong force will be needed." The governor immediately called out the entire Colorado State Militia under the command of General Brooks and ordered them to proceed to Leadville by railroad as quickly as possible.[181]

When the damage was assessed the following morning, it was found that the Coronado Mine buildings were a smoldering ruin, with an estimated $50,000 in damage to the buildings and machinery. Far worse than that, six men were dead: Bert Meir, James Benson, Andy Mahoney, Michael O'Doherty, Jerry O'Keefe and William Higgins. All except O'Keefe were union members. Others were wounded but would not give their names while being treated. When the initial three hundred state militia arrived, General Brooks posted the men at the Coronado, Emmet and Maid of Erin Mines and briefly imposed martial law on the city. The Leadville City Council met on the same day to deal with the situation, and private citizens with guns patrolled the city, holding deputy commissions from Mayor Samuel D. Nicholson and City Marshal Ernest L. Daniels. All seemed quiet until September 26, when at 3:30 a.m., shots were fired at the state militia troops guarding the city waterworks. On that day, arrest warrants were issued for twenty-five members of the Cloud City Miners' Union, including the president, vice president, secretary and executive committee members. Also arrested: Edward Boyce, president of the Western Federation of Miners. They were charged with murder but soon released for lack of evidence.[182]

If possible, tension mounted even more that same day when the first contingent of sixty-five outside strikebreakers arrived from Joplin, Missouri. These were lead miners who had been earning from $1.25 to $1.75 a day, and they must have been a lean and hungry lot. The Leadville mine owners promised them either $2.50 or $3.00 plus transportation. The state militia formed a hollow square around the Missouri men as they marched from the train depot to the Emmet Mine. The *Fort Morgan Times* reported:

> *The march was without incident until it had progressed two or three blocks along East Fifth Street. Here the popular dislike for the men who had come to replace Leadville miners came out in its full intensity. All the way from there the hissing and hooting did not cease for a moment. Women were the leaders in this chorus of contumely. At Hemlock street, where a halt was made, a stout matron with hair as white as snow, stood on the sidewalk, within six feet of the imported miners, shouting: "Oh the scabs, the hungry Missouri scabs."*

An overall Leadville view shows the devastated industrial landscape of the city in the 1890s. The pollution is still being mitigated. *Library of Congress.*

> *A younger woman in a doorway, holding up her little boy so he could see over the heads of the soldiers cried out: "See the scabs in the middle darling: when you grow up, get a gun and shoot them."*
>
> *After a brief inspection of the strangers, another shrilled-lung woman yelled: "The _____ ______ ____ haven't had a square meal for a month."*
>
> *...As a rule, the men in the crowds contented themselves with hooting and hissing. One very excitable individual announced that "if the miners were any good they'd get their guns and wipe the ___ _______ ____ scabs off the earth."*

When the formation halted in the middle of the block between Hazel and Alder Streets, a group of men pressed up against the left flank of the militia formation. When ordered to step back by Captain Grove of Company E, First Regiment, one of the men called Grove a foul name. Adjutant

A Colorado State Militiaman protects Leadville's Emmett Mine, the target of an attack by striking miners in September 1896. *History Colorado.*

General Cassius Moses, who was on horseback nearby, asked Grove, "How do you feel, Captain?" Grove answered, "Feel like fighting." Moses then replied, "Then go after him." Grove drew his saber and struck the man, John Burgeson, inflicting a wound on his cheek. The rest of the crowd fled, pursued by the militia who began smashing the men with rifle butts. Half a dozen children were knocked down and trampled in the mêlée. At least one man, a nineteen-year-old named Charles Burkard, was bayoneted in the stomach, and Michael O'Brien was "lamed" by rifle blows to the small of his back. Hundreds more Missourians poured in through 1896: 200 on October 19, 125 on October 24 and 250 on December 30, among other arrivals. Fortified stockades were built around the Little Jonny, Bison and other mines. A grand jury was empaneled to investigate the violence.[183]

On September 28, solemn funerals were held at the same church for the murdered fireman, Jerry O'Keefe, and one of the miners killed attacking the Coronado Mine, William Higgins. A procession of 1,500 union members followed Higgins's coffin to the cemetery. The miners' union made a futile attempt to stem the flow of strikebreakers into Leadville by posting a notice in other mining towns in early October. It read:

> *Leadville, Colo., Oct. 3, 1896. NOTICE: All friends of organized labor are earnestly requested to remain away from Leadville until the pending*

trouble between Cloud City Miners' Union No. 33, Western Federation of Miners, and the mine operators is adjusted. Be not deceived by false rumor that declares the strike off. Men shipped here are compelled to work inside a stockade, under guard. Will notify through circular when strike is declared off.
Edward Boyce
President, W.F. of M.[184]

It was apparent that the whole Leadville strike situation was at an impasse. Neither the union nor the mine owners and managers would budge on their demands. With massive numbers of strikebreakers being imported, the state militia protecting the strikebreakers and keeping order and some union men choosing to return to work, it seemed that the strike was losing steam. Some hope dawned in early November, when Governor McIntyre was defeated for reelection by Democrat Alva Adams. It was anticipated that he could intervene, as Governor Waite had during the Cripple Creek strike, and break the logjam.[185]

The bitterness that festered between Leadville citizens and the imported Missouri miners led to new violence on the day after Christmas when four Missouri men got into an altercation with Leadville mine lessee Patrick Carney and shot him to death in front of his own home. Carney and his wife were returning from a dance about 2:00 a.m., and Carney had been drinking. The Missourians were walking nearby. Mrs. Carney went into the house, but Carney, according to the Missourians, accosted them, threatened to arrest them for disturbing the peace and started to draw a gun. It was then that Bud Evans drew his own gun and shot Carney, killing him instantly.[186]

The violence continued into the new year when a Leadville special police officer, Henry Guyton, shot and killed a miner named Michael Douherty. Guyton and several other police were at Johnson's saloon in Leadville on January 9 along with a large crowd of perhaps one hundred people. Douherty arrived with several friends, and the police suspected him of carrying a concealed weapon. Firearms safety being what it was then, he had a large revolver stuck into the front of his trousers. When the police followed him outside and asked him about a gun, Douherty reportedly drew his pistol and fired, missing, when one of the police grabbed his arm. Guyton then drew his revolver and shot Douherty, killing him. This version was disputed by Douherty's friends, who said Guyton hit him repeatedly over the head with his pistol and then shot him in cold blood. A crowd of

Colorado State Militiamen try to relax while guarding the shaft house at the Emmet Mine in 1896. *History Colorado.*

perhaps five hundred people soon gathered outside the saloon, and the state militia provost guard was called out. They ran to the scene and cleared the street with fixed bayonets.[187]

The *Colorado Daily Chieftain* reported on January 11, "Owing to threats of lynching armed citizens turned out and reinforced the police guard at the county jail until morning, and the law and order element is on the alert tonight to prevent any trouble which may arise." A coroner's jury ruled that Guyton was justified in shooting Douherty in self-defense. It was asserted by the *Chieftain* that a dozen of the striking miners, including Douherty, were part of a conspiracy to "do up" the special police on January 9 but were surprised by Douherty's shooting and the arrival of the militia and unable to carry out their plan.[188]

The four Missourians involved in the Carney shooting and Officer Guyton were granted a change of venue and tried in Chaffee County at Buena Vista in early May. The charges against three of the Missourians were dropped, and Bud Evans was acquitted on the grounds of self-defense. Henry Guyton was also acquitted on the same grounds shortly afterward.[189]

Law enforcement in Leadville took a step forward on December 31, when the Lake County commissioners appointed Thomas F. O'Mahoney county sheriff to replace the disgraced Michael Newman. The new sheriff had served in the state legislature, was not a union member and stated that he would enforce the laws.[190]

Eugene V. Debs (1855–1926) led the nationwide railroad strike in 1894 and tried to help resolve the great 1896–97 Leadville strike. *Library of Congress.*

Labor leader Eugene V. Debs arrived in Leadville on January 13 and was given a festive reception by the union. Debs had been the president of the American Railway Union, had led the nationwide Pullman strike in 1894 and had been imprisoned for six months for violating an injunction to end the strike. He was a hero to the labor movement. Debs gave a public address at the Tabor Opera House and then met privately with the union leadership.[191]

Hopes of ending the strike peaked on January 18, 1897, when Governor Alva Adams arrived in Leadville and met with mine owners and union officials. Adams was a pro-labor Democrat who signed the bill making Labor Day a Colorado holiday and urged all working people to organize and join unions. He had served previously as governor from 1887 to 1889. The union proposed that all mines formerly paying $2.50 a day would raise wages to $2.75 and all mines currently paying $3.00 should remain at that rate. Engineers and pump men would be paid $4.00 and foremen $3.00. Strikers could be rehired provided they had not been convicted of any crime during the strike. The mine owners countered with a proposal that all mines paying $2.50 would go to $2.75 when silver prices reached $0.70 an ounce. When silver prices reached $0.75 an ounce, the pay rate would go up to $3.00 per day.[192]

The union members unanimously rejected this offer on the advice of the Western Federation of Miners president, Edward Boyce. They were perhaps buoyed by the news that $25,000 was on its way to them from WFM headquarters in Butte, Montana. Eugene Debs advised the union to accept the offer and expressed his frustration at the failure to end the strike. The *Aspen Weekly Times* reported the owners' response: "The Mine owners state

Alva Adams (1850–1922) served as Colorado's governor three times, the second time during Leadville's strike of 1896–97, which he vainly tried to resolve. *History Colorado.*

that they will pay no further attention to the union, or their ex-employees, but will operate their mines in the best way they can. All the rest of the mine pumps will be pulled immediately, including those in the upper levels. The removals of those pumps will affect not only the big mines, but a large number of the lesser."[193]

By this time, it was obvious that the energy and enthusiasm for the strike was gone. On February 5, Governor Adams recalled all the state militia in Leadville, except for a 35-man provost guard. The Western Federation of Miners headquarters in Butte sharply cut financial aid for the Cloud City Miners' Union No. 33, apparently in response to the union's refusal to accept the mine owners' offer. The result was a mass union meeting of about 1,200 members on March 9, where they voted overwhelmingly to close the strike. It appeared as if the whole effort ended in a massive defeat for the union.[194]

As Coloradans looked back in the aftermath of the strike, some good had been accomplished. An ineffectual and corrupt sheriff had been removed and replaced in Lake County. Governor Adams and the state legislature created a State Board of Arbitration that might head off future labor conflicts like the Leadville strike. But there was also plenty of recrimination to go around.[195]

Fingers pointed in every direction, with blame being heaped on the Western Federation of Miners leadership, the local miners' union, Governor McIntyre, professional agitators, the mine owners and "Leadville Johnny" Campion.[196]

Given human nature, bitterness continued after the strike was over, especially by local people against the imported Missouri strikebreakers. On March 18, "Leadville Johnny" Campion received a letter from "The Committee" demanding that he fire all non-union workers or the writers would "blow you and the Ibex off the earth." The *Herald Democrat* remarked, "For some days past a small band of agitators and their followers have been talking violence against both the managers and the non-union men

The Western Federation of Miners recruited hundreds in Silverton during the Leadville strike, illustrated by the still-standing Miners' Union Hall. *Library of Congress.*

employed on various properties. The threat is openly made that non-union men must be discharged or there will be trouble." Irish miners felt they were being blacklisted, and one of them, J.A. Brown, said, "You know damned well that for some reason the Irish members of the C.C.M.U., can't get a day's work in this damned Camp, because the sons of bitches of Mine owners have blacklisted us." Frank Moore said, "An Irishman can't get work since the strike, while the Cornish and Swedes are all working."[197]

Though the strike seemed a total defeat for the local Leadville union and the Western Federation of Miners, the Cloud City Miners' Union No. 33 survived the strike. The WFM became even more powerful after the Leadville strike, enlisting hundreds of new members in Silverton and other Colorado mining communities. WFM president Edward Boyce remained defiant, trumpeting to the WFM convention in Salt Lake City in May 1897 that union members should all arm themselves: "When ways and means are provided that every man can be furnished with the latest improved rifle...I entreat you to take action on this important question, so that in two years we can hear the inspiring music of the martial tread of 25,000 armed men in the ranks of labor."[198]

Chapter 7

NO CROWN OF THORNS

Political turmoil racked the United States in 1896. The country was amid a terrible national depression, with millions unemployed or losing their life savings, banks and businesses failing, mines closing and workers striking. There was internal dissent within both major parties, and several new parties were competing to see which one could maneuver to best advantage.

Republicans held their national convention in St. Louis in mid-June. William McKinley won the presidential nomination on the first ballot, defeating Thomas Reed and others 661½ to 204. Garret Hobart, a successful railroad attorney who was chair of the New Jersey Republican Party, was chosen as the vice presidential candidate. The Republican platform defended "sound money" and the gold standard, high protective tariffs and support for Cubans rebelling against Spanish rule. The convention's most controversial moment came when Senator Henry Teller and other Republicans favoring free silver and opposing the gold standard walked out of the convention and the Republican Party, forming their own Silver Republican Party. They denounced the Republican Party, "once the redeemer of the people, but now about to become their oppressor, unless providentially restrained by the votes of free men."[199]

Democrats were split over the free silver versus gold standard issue. A strong contingent supported President Grover Cleveland's stand supporting the gold standard and his suppression of the Pullman strike with federal

At the 1896 Republican Convention in St. Louis, William McKinley easily won nomination for president on the first ballot. *Library of Congress.*

troops, while many Democrats hated Cleveland for those same reasons. The blame for the Panic of 1893 and the Silver Crash was put squarely on Grover Cleveland and his supporters by silver advocates, who were obviously gaining strength in state and local elections. The dilemma for the Democrats was: who would replace Cleveland and Vice President Adlai Stevenson as the party's 1896 candidates?[200]

One-eyed governor "Pitchfork Ben" Tillman of South Carolina (so called because he threatened to stick a pitchfork in Grover Cleveland's ribs) was a supporter of bimetallism and admired by many Populists. Yet his wild, intemperate rhetoric, rampant racism and praise of the Confederacy were liabilities that barred him from nationwide popularity.

Former governor Horace Boies of Iowa was another strong advocate for bimetallism. As the only Democrat to be voted in as Iowa's governor from 1855 to 1933, even the *New York Times* lauded Boies in 1893 in blazing headlines praising Boies's "sterling qualities." Boies had been strongly boosted by silver advocates as the party's presidential candidate in 1892.

Though Benjamin "Pitchfork Ben" Tillman (1847–1918) supported free silver, his virulent racist and secessionist rhetoric barred him from national candidacy. *Library of Congress.*

Congressman Richard P. Bland of Missouri was a great hero of silver advocates for his passage of the Bland-Allison Act of 1878, restoring bimetallism and requiring the treasury to buy $2 to $4 million in silver each month. As chairman of the Committees on Mines and Mining and Coinage, Weights and Measures, Bland had a huge influence on United States monetary policy. Of all the possible candidates, Bland had the strongest support going into the convention.

Supporters of the gold standard boosted John Carlisle, Cleveland's secretary of the treasury.[201]

When the Democrats met in Chicago, it seemed that none of the prospective candidates could win the requisite two-thirds majority of delegates. The platform committee, dominated by free silver advocates, created a platform supporting free coinage of silver at sixteen to one, government regulation of large corporations, the workers' right to join unions, a graduated income tax and direct election of senators by the people instead of state legislatures.[202]

Silver delegates were deadlocked between Boies and Bland. And there was some notice of a young Democratic ex-congressman from Nebraska who sprang to national fame for his outstanding oratory opposing the McKinley Tariff in Congress: William Jennings Bryan. He had recently made an unsuccessful bid for a seat in the U.S. Senate but had circulated throughout the country speaking in support of bimetallism. Many of his utterances coincided with positions of the Populist Party.[203]

Working on the resolutions committee, Bryan had been unable to speak until the second day of the convention, when delegates were weary of the infighting, and luckily, he could follow two lackluster speakers supporting the gold standard. When asked by a reporter who the nominee would be, Bryan replied, "Confidentially speaking, I will be." The reporter took the remark for a joke. To another confidant, Bryan astutely described his nomination as the "logic of the situation."[204]

When Bryan was called on to speak, he bounded up the stairs to the speaker's platform two at a time, then stood in a dramatic pose with one arm upraised to quiet the cheering crowd. He began with an idealistic

The 1896 Democratic National Convention was electrified by William Jennings Bryan's "Cross of Gold" speech, leading to his nomination for president. *Library of Congress.*

proclamation: "The humblest citizen in all the land, when clad in the armor of a righteous cause, is as strong as all the whole hosts of error that they can bring. I come to you in a cause as holy as the cause of liberty: the cause of humanity." But then he went on the attack, pointing out that the majority of Democrats were in favor of free silver: "Our silver Democrats went forth from victory to victory, until they are assembled now, not to discuss, not to debate, but to enter up the judgement rendered by the plain people of this country." Then Bryan emphasized that the working-class people of America were equal in dignity to the upper classes and went on to make a rhapsodic plea for the people of the western prairies, his own people:

> *Ah, my friends, we say not one word against those who live upon the Atlantic coast, but those hardy pioneers who braved all the dangers of the wilderness, who have made the desert to blossom as the rose—those pioneers away out there, rearing their children near to nature's heart, where they can*

mingle their voices with the voices of the birds—out there where they have erected schoolhouses for the education of their children and churches where they praise their Creator and the cemeteries where sleep the ashes of their dead—are as deserving of the consideration of this party as any people in this country.

Defending the legitimacy and legality of the progressive income tax, demanding that only the federal government and not banks issue currency, Bryan denounced the concept of waiting for an international agreement to adopt bimetallism. Then he declared the "line of battle":

It is simply a question that we shall decide upon which side shall the Democratic Party fight. Upon the side of the idle holders of capital, or upon the side of the struggling masses? That is the question that the party must answer first, and then it must be answered by each individual hereafter. The sympathies of the Democratic Party, as described in the platform, are on the side of the struggling masses, who have ever been the foundation of the Democratic Party.

Then Bryan denounced the "trickle-down" theory of economics:

There are two ideas of government. There are those who believe that if you just legislate to make the well-to-do prosperous, that their prosperity will leak through on those below. The Democratic idea has been that if you legislate to make the masses prosperous their prosperity will find its way up and through every class that rests upon it.

In the final part of the speech, Bryan turned angry, defying the urban East and asserting that bimetallism was synonymous with American independence:

I tell you that the great cities rest upon those broad and fertile prairies. Burn down your cities and leave our farms, and your cities will spring up again as if by magic. But destroy our farms and the grass will grow in the streets of every city in the country.

It is the issue of 1776 over again. Our ancestors, when but 3 million, had the courage to declare their political independence of every other nation upon earth. Shall we, their descendants, when we have grown to 70 million, declare that we are less independent than our forefathers?

And at the very end, Bryan dramatically acted out his immortal peroration, grasping an imaginary crown of thorns and placing it on his head, then spreading out his arms in a gesture of crucifixion:

> *If they dare to come out in the open field and defend the gold standard as a good thing, we shall fight them to the uttermost, having behind us the producing masses of the nation and the world. Having behind us the commercial interests and the laboring interests and all the toiling masses, we shall answer their demands for a gold standard by saying to them: you shall not press down upon the brow of labor this crown of thorns, you shall not crucify mankind upon a cross of gold.*[205]

As Bryan stepped back, there was dead silence for a few seconds. Then—cacophony! A deafening wave of sound combined shouting, cheering, weeping, stomping, singing—every noise that the crowd could produce resounded in a chaotic medley. Parades of Bryan supporters snaked through the aisles. Fistfights broke out as Bryan men tried to grab banners of the eastern delegations. The crazed spectacle continued for a solid forty-five minutes, the longest and loudest demonstration ever held at any political convention.[206]

Just what was it that caused such a violent reaction among the Democrats? The answer is that Bryan put into eloquent words what others had been feeling for decades but struggled to express. Defense of free silver, placing the working-class people of the United States on an equal level with the capitalists and bankers, the heartfelt tribute to the people of the prairies and plains and the defiant biblical conclusion with its vivid imagery—these expressed the entire world view of most Democrats and Populists in 1896. Add to that Bryan's forceful, sincere and powerful delivery and the effectiveness of the speech can be gauged. To this day, it is considered one of the most influential political speeches in American history.[207]

William Jennings Bryan (1860–1925) energized audiences on the 1896 campaign trail with his strong, magnetic personality and fiery oratory. *Library of Congress.*

The myth is that the "Cross of Gold" speech catapulted Bryan into the Democratic presidential nomination. The truth is less dramatic. Bland supporters were reluctant to surrender, and it was only during the fifth ballot that Bryan overcame opposition and was unanimously declared the Democratic nominee. Arthur Sewall, a wealthy shipbuilder, railroad president and banker from Maine, was chosen as Bryan's running mate to appease easterners suspicious of Bryan's Populist leanings. Democrats who supported "sound money" and the gold standard bolted from the party, forming their own Gold Democrat organization. When they met in Indianapolis in early September, they nominated Senator John Palmer of Illinois for president and former Confederate general Simon Bolivar Buckner of Kentucky for vice president.[208]

The Populist Party was fighting to retain its influence in 1896, but the party's stand against the railroads and large corporations on behalf of struggling farmers and silver miners brought it a hard core of support. The biggest issue faced by the Populists was leadership. Who could offer the best chance of overcoming regional rivalries to appeal to voters on a national level? Another crucial concern: should the Populists join the Democrats in a "fusion" presidential ticket or still insist on fielding their own candidates? Warily, the Populists scheduled their convention after the other two, in late July in St. Louis. That way, they could evaluate the Democratic candidates and decide whether they wanted a "fusion" ticket. The Populists did not lack for possible presidential candidates.[209]

First on most people's list was James Weaver, who had run for president as a Populist in 1892, racking up an impressive number of votes and carrying four western states.

Tom Watson of Georgia was another former congressman who was originally a Farmers' Alliance candidate. Watson had maneuvered the bill for Rural Free Delivery through Congress, providing an enormous boon for farmers in remote areas.

"Sockless" Jerry Simpson was a Kansas farmer and rancher who moved from the Union Labor Party to the Populists. When he ran for Congress against railroad attorney J.R. Hallowell, Simpson attacked Hallowell for wearing silk stockings. Hallowell countered that he would rather wear silk stockings than none, and this earned Simpson his enduring nickname: "Sockless" Jerry. He won the election. In 1896, Simpson was running to regain his seat in Congress after losing in 1894.

When the Populists met in St. Louis, there was a bitter struggle between those who backed a "fusion" ticket, supporting Bryan and Sewall, and "middle of the road" Populists who wanted independent Populist candidates. Fistfights broke out, and pistols were drawn. In the end, there was a compromise solution, with the convention voting to support Bryan

Left: "Sockless Jerry" Simpson (1842–1905), a struggling Kansas farmer and town marshal, became one of the Populist Party's strongest congressional leaders. *Library of Congress.*

Right: Tom Watson (1856–1922) was the Populist Party vice presidential candidate, saddling William Jennings Bryan with separate Democratic and Populist running mates. *Library of Congress.*

for president, but with a separate Populist vice presidential candidate, Tom Watson. This decision was to induce much ridicule against Bryan and was an albatross around his neck during the campaign.[210]

The Silver Republicans and National Silver Party met in St. Louis at the same time as the Populists, and both named Bryan and Sewall as their presidential and vice presidential candidates. When Judge C.K. Scott of Omaha was called to the podium at the Silver Republican convention, he said, "Oh God send pestilence and disease and vermin and war and famine among us if you will, but in thy good providence, Oh God, deliver us from another four years of oppression under Grover Cleveland."[211]

The two presidential candidates were similar in many ways. Both were Midwesterners, both began as attorneys and then advanced into politics, both were devoutly religious and devoted to their families and both were supporters of bimetallism, McKinley only moving away from it reluctantly.

Former congressman and governor of Ohio William McKinley was known as "The Major" from his service in the Civil War. After the war, McKinley settled into law practice in Canton, Ohio, and was elected to Congress in 1877. He voted for the Bland-Allison Act and restoration of bimetallism in 1878. McKinley became the Republican Party's leading exponent of protective tariffs, stating that they were "the law of nature, the law of self-preservation." With his strong support of protective tariffs, McKinley was known as the "Napoleon of Protection" and was often depicted by cartoonists in a Napoleonic uniform and hat.

McKinley adopted a compromise position on free silver, stating that bimetallism might be adopted after an international agreement was reached. He only moved to total support of the gold standard when forced to do so by the 1896 Republican platform.

McKinley's wife, Ida, suffered from epilepsy, and at times she would suffer a seizure at a public gathering. He kept a large white handkerchief handy, which he would place over her contorted features until the seizure passed and continue as if nothing had happened. His devotion to her was total.[212]

William Jennings Bryan was a child of the prairies, born in Illinois but later moving to Nebraska. He and his family were devout Presbyterians. Though depicted as a barnyard hayseed by Republicans, Bryan received a solid grounding in Greek and Latin, and he excelled in rhetoric and oratory while in college. Beginning as an attorney in Lincoln, Bryan was elected to Congress as a Democrat and defender of farmers' and workers' rights in 1890. His greatest asset as a politician was his wife, Mary, who was also a well-educated attorney with shrewd political sense. When elected to Congress, Bryan struck out boldly in defiance of the McKinley Tariff and on behalf of bimetallism. Known increasingly across the country for his oratorical skill, Bryan combined homely metaphors with biblical quotations and allusions. Running for the Senate in 1894, Bryan was defeated because of the backlash against the Democrats caused by the Panic of 1893.[213]

If Bryan's shrewdest political adviser was his wife, Mary, McKinley's was his best friend, Marcus Alonzo "Mark" Hanna. Mark Hanna was a wealthy Cleveland businessman with wide-ranging interests in coal, iron, railroads and Great Lakes shipbuilding. He had been involved in presidential campaigns beginning with his backing of James Garfield in the 1880 election. He became a strong gold standard supporter and an admirer and extremely close friend of William McKinley, who he considered to be a political genius. Hanna became perhaps the first true modern presidential campaign manager during the 1896 campaign. He was routinely depicted by Hearst

A *Puck* cartoon shows McKinley, the "Napoleon of Protection," tugged by both silver and gold advocates while straddling the silver issue. *Library of Congress*.

Left: After the deaths of Ida and William McKinley's two young daughters, Ida McKinley (1847–1907) became a semi-invalid, afflicted with epilepsy. *Library of Congress*.

Right: Mark Hanna (1837–1904) was William McKinley's closest friend and campaign manager, guiding the Republicans to victory in the 1896 presidential campaign. *Library of Congress*.

cartoonist Homer Davenport as a bloated, swollen plutocrat wearing a suit emblazoned with dollar signs. Hanna was shrewd, resourceful and above all utterly loyal to McKinley. He was big, blunt, gruff and decisive.[214]

Bryan's nomination caught Hanna and McKinley by surprise. They expected the tariff to be the main campaign issue and to face an older, less energetic rival like Richard Bland. They were forced to switch tactics and focus on the monetary issue and a younger candidate with inexhaustible energy and an enormously enthusiastic following. When it became obvious that Bryan was going to crisscross the country by rail, speaking at every opportunity, Hanna urged McKinley to also barnstorm on his own behalf. McKinley wisely demurred, saying, "You know I have the greatest respect for your wishes, Mark, but I cannot take the stump against that man....I can't outdo him, and I am not going to try."[215]

The result was McKinley's famous "front porch campaign." While Bryan strenuously traveled the country by rail, giving speeches everywhere, McKinley stayed home in Canton, Ohio, greeting Republican delegations

William McKinley greets a Republic delegation visiting the McKinley home during the 1896 presidential campaign. *Wikipedia.*

and giving gracious remarks on his front porch. Railroad companies offered reduced ticket rates to Canton, prompting the comment that visiting McKinley was "cheaper than staying at home." Thus, McKinley cultivated an image of quiet, respectable domesticity. Visiting delegations often brought unexpected gifts, such as a pair of raccoons from Kentucky and a pair of bald eagles from Montana.[216]

Colorado's newspapers almost unanimously supported Bryan and free silver in overwhelmingly enthusiastic terms. The *Daily Mining Record* ran a bit of doggerel:

Bryan, Bryan.
McKinley's a cryin'.
There's no use tryin'.
The eagle's flyin'.
They can't beat Bryan.
B-R-Y-A-N – B-R-Y-A-N[217]

The *Grand Junction Weekly Star-Times* asserted: "There is one political creed in Colorado, this year of our Lord—'Free Silver.' A few there are who are out for chunks of pie, who are traitors to this state, but the people have got them on their list and the day of reckoning is nigh at hand."

The *Boulder Daily Camera* went overboard in its praise of the Democratic platform: "Boulder citizens, regardless of politics, seem to be one in expressing the sentiment that no greater document than the national democratic platform has ever been delivered from any body in this country since Thomas Jefferson gave the declaration of independence to the world."

The *Aspen Weekly Times* trumpeted: "Chosen by the people as their strongest candidate, he is honored as few men are honored. His is the glorious task of leading the people in the struggle for their rights."

The *Rocky Mountain Sun* saw Bryan as a savior for the people of Colorado:

> *The star of redemption for an oppressed people has indeed arisen in the person of William Jennings Bryan and like that fearless monarch of the air the American eagle who soars unflinchingly into the very eye of the sun, so will the people of this nation march to the polls on the third of November, and with eyes firmly set upon their guiding star, swerving neither to the right or to the left, will press on to their duty and cast their ballot for the rehabilitation of the cause of silver.*[218]

Approval of Bryan was not unanimous in Colorado. The *Chaffee County Republican* condemned Bryan as vindictive, saying that he was "the first candidate for the presidency who has pronounced that state of New York an enemy's country, and he is the first that has even 'branded' people who did not support him as traitors." The *Fort Morgan Times* accused Bryan of being an enemy to labor: "If elected he will lend more of his energies to a free trade law than the furtherance of the interest of silver."[219]

Lingering bitterness from the Civil War was much more visceral. A member of the Lincoln Republican League of Silver Plume wrote in *The Woman Voter and Western Woman* that she could never vote for a Democrat: "Why I am not a woman follower of the Democratic Party? Because that flag I was taught to love, and cherish, that flag, our pride our boast, my emblem of nationality, and the record of our glory, was assailed by traitorous hands, and trailed in the dust, at the bidding of the Democratic party."[220]

Colorado journalists typically excoriated McKinley as a mere tool of Hanna and the trusts. The *Fort Collins Courier* asserted that "the trusts are pouring out money like water to elect McKinley, but they calculate to repay

William McKinley (1843–1901) battled to win the 1896 Republican nomination as a reformer with the slogan "The People Not the Bosses." *Library of Congress.*

themselves by robbing the people. It would be strange, however, if the people forged their own chains by selecting the double faced demagogue to the presidency."

In another article, the *Courier* noted, "Mark Hanna, his chief backer, is the embodiment of the trust idea. Does it not look as if the government, in case Mr. McKinley should be elected, would be administered by the trusts for the trusts and with the trusts?"

The *Aspen Tribune* accused the Republicans of "bulldozing" company employees to join McKinley clubs, citing several men who had been fired for refusing to join. The *Tribune* also asserted that "a number of the railway corporations have detectives whose business it is to circulate among employees and 'spot' those who are talking for Bryan and free silver."[221]

Bryan's momentum grew through the summer as he campaigned in the Great Lakes states and through the South. His campaign schedule was incredibly strenuous. In one day, August 26, on his way to Chicago from New York, he spoke to about forty thousand people in groups as large as ten thousand. In Springfield, Ohio, a major farm machinery manufacturing center, ten thousand people gathered to hear him speak. In Fredericksburg, Virginia, he was given a "huge" reception. The *Colorado Daily Chieftain* noted that during Bryan's September swing through the South, in North Carolina:

> *The train whizzed by the little village of Marshall and the 200 people on the depot platform shouted as they caught a glimpse of Mr. Bryan, who came to the back platform of the train. By the track alongside the mountain barefooted men grouped themselves in knots, and yelled excitedly when they saw the banners displayed on the sides of the cars which read: "We carry the next president, William Jennings Bryan."*[222]

But by then, Mark Hanna had cracked open the coffers of eastern capitalists such as his fellow Buckeye John D. Rockefeller. Rockefeller's Standard Oil donated $250,000 to the McKinley campaign. Others such as J.P. Morgan followed suit until a war chest of about $3.5 million was assembled, not counting other contributions at the state and local level. In

William Jennings Bryan (*far left*) and his wife, Mary Baird Bryan (*far right*), during an 1896 whistle stop at Westerville, Ohio. *From* The First Battle.

Paul Koenig's excellent political biography of Bryan, he estimated that the total Republican expenditures may have been as high as $16 million, a figure disputed as too high by Karl Rove in his book on the campaign. There were also many in-kind contributions such as reduced railroad fares to Canton, Ohio, to visit McKinley.[223]

The Republican "literary bureau" distributed about 250 million lithographs, posters, song books and pamphlets supporting McKinley and Hobart. Hanna also hired 1,400 speakers, including young Theodore Roosevelt, to fan out across the country denouncing Bryan and praising McKinley and his policies. Roosevelt remarked of Hanna, "He has advertised McKinley as if he were a patent medicine!" Roosevelt's speeches denounced Democratic and Populist politicians as "vicious mob-leaders" and their supporters as "men of little intelligence."[224]

In contrast, the Democratic Party chairman, Senator James K. Jones of Arkansas, was a lackluster fundraiser, and the Democrats only gathered around $400,000 during the entire campaign. The best support for the silver cause came from the publications of William H. "Coin" Harvey, a monetary

Young Theodore Roosevelt campaigned for McKinley, equaling Bryan's oratorical skills while denouncing Democrats and Populists as either dim-witted or radical revolutionaries. *Library of Congress.*

expert who wrote the enormously popular book *Coin's Financial School.* In it, he explained the various aspects of the monetary question, coming down squarely on the side of bimetallism. Quixotic newspaper magnate William Randolph Hearst was another strong supporter, gathering some $40,000 in contributions, half of it from Hearst himself. Citizens of Aspen, Colorado, gathered $35,000 in donations to the Bryan campaign.[225]

As the campaign of 1896 ended, the *Boulder Daily Camera* offered this assessment:

> *The most magnificent campaign ever waged in this country closed at Omaha last night, when William J. Bryan, the nominee of three parties, addressed five magnificent assemblies. For three months the citizens of two continents have noted with amazement the nerve, brilliancy and solid attainments of this wonderful product of the West. An almost unheralded messenger of a cause, he has made himself within that period the recognized chieftain of debate and apostle of human liberty. He will be opposed, his ideas scoffed at and his cause condemned, but no honest opposition will withhold from William Jennings Bryan the plaudits due a man of matchless nerve, unequalled eloquence and supreme, almost divine, devotion to a cause.*[226]

Despite Bryan's strenuous campaigning and the fact that he was supported by the majority of voters in the West and South, the decisive election victory on November 3, 1896, went to William McKinley, Garrett Hobart and the Republican Party. The vote totals were: 7,112,138 for McKinley, 6,510,807 for Bryan. The Electoral College total was much more lopsided, since McKinley won 271 votes (60.63 percent) and Bryan 176 (39.37 percent). Though Bryan's support was more widespread, McKinley won the most populous regions of the country: New England, the Mid-Atlantic states and the Midwest's older states.[227]

Bryan won Colorado with 84.96 percent of the votes to McKinley's 13.86 percent; 189,539 Coloradans voted. A change of less than 5 percent of the votes would have gained Bryan the victory in eight states.[228] As the dust settled, Coloradans were forced to deal with the defeat of their silver champion William Jennings Bryan, an uncertain future for the silver mining industry and the lingering ordeal of the Silver Crash. Millions of Americans lamented along with Vachel Lindsay in his poem "Bryan, Bryan, Bryan, Bryan":

> *I brag and chant of Bryan, Bryan, Bryan,*
> *Candidate for president who sketched a silver Zion.*
> *Election night at midnight:*
> *Boy Bryan's defeat*
> *Defeat of western silver,*
> *Defeat of the wheat.*[229]

Chapter 8

PEACE AND WAR

Naturally, there was bitter disappointment at the defeat of the free silver cause and William Jennings Bryan. But in general, there was surprisingly little resentment and recrimination. Indeed, Bryan and his Democratic, Populist and Silver Republican supporters could be proud. During the campaign, Bryan traveled some eighteen thousand miles by rail. He personally gave more than three thousand speeches and gained some 900,000 more votes than Grover Cleveland had as a Democratic candidate in 1892. His defense of free silver gained him enormous popularity throughout the western mining states. In Aspen, an astonishing 99 percent of the voters supported Bryan.[230]

When Bryan visited Colorado in late November, his journey had the aura of a winning candidate rather than a dejected loser. When he reached Denver on November 24, he spoke to more than five thousand people at the train station, and the reaction of the Denver citizens was "one continued ovation....The people clung to Bryan's carriage and followed it from corner to corner and street to street." At Colorado Springs, he spoke from the balcony of his hotel, saying, "You act as though you were in the midst of a campaign or were celebrating a victory just won. You are right; we are in the midst of a campaign. We are going to accomplish in 1900 what we failed to accomplish in 1896."[231]

As Republicans celebrated their victory, they, too, could point with pride to their accomplishments. The "front porch" campaign strategy succeeded brilliantly, enabling McKinley supporters to meet their candidate in a quiet, intimate, domestic setting rather than in a crowded, noisy public venue. Republican employers undoubtedly used threats of shutdowns and firings to coerce their employees into voting for McKinley, but these actions were at least equaled by suppression of African American Republican votes in southern states dominated by Democratic political machines.[232]

In retrospect, the 1896 presidential campaign was one of the pivotal political battles in American history. Despite his defeat in 1896 and two other presidential elections, William Jennings Bryan was still a vital transitional figure in American political life. As Margaret Leech noted, "Bryan was of service to his country in laying bare the abuses of concentrated wealth and its control of government. He touched the laggard conscience of America, and disturbed its complacent absorption in material success." William McKinley was also a decisive figure, as the alliance between the Republican Party and the financial leadership of the country established during his presidency resulted in the election of six Republican presidents between 1896 and 1928.[233]

Those Coloradans who hoped to revive silver mining received a terrible shock as a second Silver Crash hit the already suffering silver industry. In May 1897, international silver prices suddenly dropped dramatically. From an average of 67 cents an ounce in April, they fell to 60 cents by late May and then even further to 43 6/10 cents an ounce in August. It was feared that mines and smelters would have to close if the price dropped any further. Apparently, the cause of the crash was Japan going on the gold standard, combined with Peru prohibiting importation of silver, instead using its own domestic silver to coin money. The *Silver Cliff Rustler* recommended that all silver producers join in a gigantic "silver trust" and demand that silver be sold at no less than $1.00 an ounce. Silver prices began to inch upward and by the end of 1897 were at about 58 cents an ounce—a far cry from 83 cents an ounce in January 1893. In contrast, gold mining was flourishing, illustrated by the fact that six times the amount of Colorado gold was being produced in 1897 compared to 1892.[234]

At the 1896 Republican convention, silver advocates led by Senator Henry Teller walked out of the convention and organized the national Silver Republican Party, supporting Bryan and Sewall in the presidential election. After Bryan's defeat, the Silver Republican Party within Colorado split bitterly over reuniting with the "McKinleyite" national Republican Party—a split that would become so virulent it would lead to a riot, a gun battle and death.

Henry Wolcott (1846–1921) ran for Colorado governor in 1898, hoping to unite the Silver Republican and national Republican Parties. *History Colorado.*

During the 1896 election, Senator Edward Wolcott, who had been a staunch silver advocate, stayed within the national Republican Party and did not split off to join the Silver Republicans. President McKinley appointed him as part of a delegation visiting various European countries in 1897 to discuss a possible international agreement on bimetallism—an inconclusive effort. Upon his return, Wolcott and his brother Henry engaged in a concerted drive within Colorado to reunite the Silver Republicans and the national Republican Party.[235]

In 1897, an election to choose the chief justice of the Colorado Supreme Court loomed in November. Judge Charles Hayt was selected by the Silver Republicans as a candidate for the office. In turn, the national Republican Party also chose Judge Hayt as their candidate—a decision seen as an attempt to co-opt the Silver Republicans and unite the two parties. The Leadville *Herald Democrat* spluttered somewhat incoherently: "The appearance of decay of the silver cause, which would be claimed by the eastern gold men on the strength of Judge Hay's reelection by a union of McKinleyites and Silver Republicans: that they—the gold men—would promptly seize upon this as evidence of the restoration of Republican harmony in Colorado, meaning, of course, the supremacy of the followers of the administration."[236]

The election was confused and chaotic. The Democrats selected Judge John A. Gordon to run, but he withdrew in favor of Populist judge William H. Gabbert. Gabbert then ran in some counties on a fusion ticket of Democrats, Populists and "breakaway" Silver Republicans. In some counties, he ran on separate party tickets. The *Colorado Transcript* commented on the "tangle of fusions" in various counties. Hayt ran on the Republican and "mainstream" Silver Republican tickets. Gabbert won by about three thousand votes, a solid majority.[237]

In 1898, the split within the two Silver Republican factions widened. A legal battle over the Silver Republican name and emblem—a dove—resulted in victory in court for the faction favoring reunion with the national Republicans, led by Richard Broad Jr. In response, the faction led by James

H. Blood, favoring an independent Silver Republican Party, filed for official designation as the Teller Silver Republican Party. On September 7, the Teller Silver Republicans met in convention at the Colorado Springs opera house. The Broad faction of the Silver Republicans attempted to take over the building, claiming that they had prior right and had rented it. When the Teller people refused to leave, the building was stormed in a hail of gunfire by seventy-five of the ever-present thugs from Denver, led by the El Paso county sheriff and Colorado Springs police chief. One of the Teller Silver Republicans, Charles Harris of Denver, was shot and killed and another man seriously wounded. Fierce accusations from both sides followed the riot.[238]

The Wolcott brothers took advantage of the turmoil within the Silver Republicans to draw the "mainstream" Broad faction even closer to the national Republican Party. They boosted Pueblo's smelter king, Simon Guggenheim, as a candidate for governor. Guggenheim was an attractive candidate primarily because of his wealth. The "mainstream" Broad Silver Republicans did indeed nominate Guggenheim as their gubernatorial candidate, but he withdrew less than a month later. With the election looming shortly, choosing a gubernatorial candidate quickly was essential. Henry Wolcott stepped in and filled the gap, in a possibly prearranged maneuver. Wolcott was also running as a national Republican, so the maneuvers of the Wolcott brothers and the alliance between national Republicans and "mainstream" Broad Silver Republicans were complete. Most Colorado newspapers skewered Senator Wolcott, calling for his resignation. The Leadville *Herald Democrat* called him a "political burglar" and "a pariah in Colorado politics, a skulking political coyotte [*sic*] a trespasser, a larcenist."[239]

Simon Guggenheim (1867–1941) and his family dominated Colorado's smelting industry. A leading Republican politician, he was elected as a U.S. senator in 1907. *Library of Congress.*

The bloody rivalry among Colorado political factions was overshadowed in 1898 by a military conflict: the Spanish-American War. In Cuba and the Philippines, independence movements resulted in years of guerrilla warfare and brutal, repressive responses from the Spanish army, including herding local people into concentration camps in Cuba. In the United States, there was widespread sympathy with the Cuban rebels and pressure on President McKinley to intervene either diplomatically or militarily. The *Colorado Daily Chieftain* suggested that American sugar companies could simply buy Cuba from Spain and operate it as a giant sugar plantation. The *Colorado Transcript* noted, "The interests of Americans in Cuba and our commerce with that island will not permit us to stand by and see the revolution go on indefinitely." The Hearst and Pulitzer newspapers were virulently stoking war fever in the United States with sensational (and sometimes fictional) atrocity stories and illustrations condemning the Spanish actions in Cuba.[240]

On February 15, 1898, the U.S. Navy cruiser *Maine*, which had been dispatched to Cuba to protect American citizens and property amid the Cuban revolution, blew up and sank in Havana Harbor with the loss of 266 American sailors. Although the real cause of the explosion has never been definitively proven, American public opinion overwhelmingly blamed Spain, and war became inevitable. By April 1898, the *Telluride Daily Journal* was declaiming, "It is now reasonably certain that the United States must go to war with Spain." The *Fairplay Flume* agreed, printing a belligerent poem based on an old British music hall song from the 1870s:

We don't like to fight,
But by jingo if we do,
We've got the ships,
We've got the land,
And we've got the money too.
We don't like to war,
But by jingo if we do,
O'er Havana will float
The red, white and blue.[241]

The Spanish government declared war on the United States on April 24 in response to an American government ultimatum to evacuate Spanish forces from Cuba, and the United States government reciprocated the next day. Coloradans enthusiastically responded to the declaration of war. The *Glenwood Post* asserted, "There will never be an armed intervention in Cuba by consent of all the people, but three-quarters of them want it, and want it now."[242]

The Colorado State Militia—renamed the Colorado National Guard by the state assembly in 1897—was not the first military force to leave Colorado for the war. The regular army soldiers of the Seventh U.S. Infantry and

USS *Maine* enters Havana Harbor, where it blew up on February 15, 1898. The explosion killed 266 U.S. sailors, igniting the Spanish-American War. *Wikipedia.*

Second U.S. Cavalry, stationed at Fort Logan, a few miles south of Denver, were the first to disembark for Cuba. They received orders to entrain for the East Coast even before war was declared, on April 18. The Seventh Infantry marched through Denver to Union Station on April 20 with more than 100,000 people cheering them on. The Chaffee Light Artillery fired a twenty-one-gun salute, and the *Aspen Tribune* reported that "cannon, revolvers, bombs and steam whistles kept up a din greater than ever before heard in this part of the country." When the African American Buffalo Soldiers of the Twenty-Fourth U.S. Infantry stopped at Union Depot in Pueblo on April 21 on their way south, they were given an equally enthusiastic reception.[243]

Colorado's Silver Crash politics entered even into preparations for the Spanish-American War. Local newspapers expressed impatience at the slow war progress, blaming the situation on interference by international speculators in Spanish bonds, war profiteers and Mark Hanna. The *Aspen Tribune* complained, "As yet not a single shot has been fired in Cuba not a soldier lodged there….Must we wait for the bond syndicate and the commissary contractors to get in their fine work before they begin to saw wood." The *Colorado Transcript* snarled in June, "With no obligation to Mark

The mast of the USS *Maine* was later installed at Arlington National Cemetery as a memorial to the sailors who died. *Wikipedia.*

Hanna, no $16,000,000 corruption fund raised from corporations for the last presidential campaign, Wall Street Stock Ticker, no Tory holders of Spanish bonds, this war would not have lasted thirty days."[244]

The Colorado National Guard was ordered to assemble in Denver on April 30, and more than one thousand men responded. Their encampment, just east of City Park, was dubbed Camp Alva Adams after the governor.

The *Walsenburg World* printed an engaging account of the soldiers' first night in camp:

> *As it was raining when the first troops arrived, they reached the grounds covered with mud from head to foot and their clothes were soaking wet.... However, the boys enjoyed it. It was their first taste of soldier life and many*

> *of them were still so excited that they could not have slept even if a soldier's bed was made of down. All classes of songs were sung. "Yankee Doodle" and "Nearer My God to Thee" seemed to be the favorites. In one tent "A Hot Time in Old Town" was repeated a score of times.*[245]

U.S. Secretary of War Russell Alger decided on quotas of troops to be requested from each state, based on population. The Colorado quota was one infantry regiment and two cavalry troops. The infantry unit, dubbed the First Colorado Volunteer Infantry, was reorganized from the two Colorado National Guard regiments. Luckily, the Colorado National Guard was fully supplied with uniforms and equipment and had been on active duty several times in recent years because of deployment in the silver mining strikes.[246]

It was assumed that all Colorado's troops would be sent to Cuba, and that was the theme of "Colorado's Battle Song" sung throughout the state in school and churches to the tune of "Marching Through Georgia":

We're soldiers from the Silver State
Centennial boys are we—
We're children of Old Glory
From a mile above the sea
And all the sons of Cuba's star
Forever shall be free
As we go marching to Cuba!
Chorus:
Hurrah! Hurrah! We won't forget the Maine*!*
Goodbye! Goodbye! Old yellow rag of Spain!
Let us swell the chorus from the
Mountain and the plain
As we go marching to Cuba!

Two more fiercely warlike verses followed.[247]

As the Colorado National Guard assembled in Denver, Admiral George Dewey's Pacific Fleet annihilated the Spanish fleet in the Battle of Manila Bay on May 1. With the battle fleets of other nations hovering in Manila Bay, anxious to pounce and take control of the Philippines, the McKinley administration decided to send American units there to seize the islands from Spain. Instead of being transported to Cuba, the First Colorado Volunteer Infantry was ordered to board trains to San Francisco and then travel by ship to Hawaii and the Philippines.[248]

When the First Colorado marched to Union Station on May 17 to depart, the *Fowler Tribune* described the scene:

> *That the soldiers were volunteers could be readily understood from the people who were at the depot. The governor and his wife, the society cult, the business man, and the professional man greeted the soldiers and bade them godspeed. Often the tears were wiped away with a lace handkerchief, and baskets of chicken and possibly champagne loaded the seats of the train.... The largest multitude of people that has ever gathered at the depot was present to see the soldiers depart, and the cheering was tremendous....Some of the women grew hysterical, and one or two fainted and many wept silently. The soldiers bore up well. A cinder or two from the smoke-stack of an engine got into the eyes once in a while, but that was all.*[249]

On May 30, the two troops of Colorado cavalry left for Fort D.A. Russell near Cheyenne, Wyoming, where they were consolidated along with troops from other states into the Second U.S. Volunteer Cavalry Regiment. This unit was referred to in Colorado newspapers as the "Cowboy Cavalry"

Soldiers of the First Colorado Volunteers leave Denver in 1898 on their way to San Francisco and ultimately the Philippines. *History Colorado.*

regiment, but it was usually dubbed "Torrey's Rough Riders" after its commander, Judge Jay L. Torrey from Wyoming. The unit was sent to Jacksonville, Florida, where it stayed during the rest of the short war, never seeing action.[250]

The First Colorado departed for the Philippines on the transport ship *China*. On the way, the ship stopped in Hawaii, Guam and Wake Island, where a few troops rowed ashore and took possession of the uninhabited island for the United States. While in the Philippines, the Coloradans fought well and had the honor of being the first to storm the Spanish defenses of Manila and plant the American flag on the fortifications. After it became clear that the United States government was planning to annex the islands, the Philippine rebels against Spain turned from allies into enemies, and the war known as the Philippine Insurrection began. Though the war with Spain ended on August 12, 1898, the Coloradans remained stationed in the Philippines fighting the Filipinos until they returned home in September 1899.[251]

American troops—possibly the First Colorado Volunteers—atop Spanish defenses after capturing Manila, capital of the Philippines, during the Spanish-American War, 1898. *Library of Congress.*

Even during the Spanish-American War, the unity shown by Coloradans in support of their troops did not extend to politics at home. Though the shooting war with Spain was over quickly, the political war to defend silver continued in Colorado and grew in intensity as the midterm 1898 elections loomed. Roughly 12 percent of American workers were still unemployed in 1898, while the unemployment rate in Colorado was much higher at an estimated 25 percent. Many Coloradans still saw free coinage of silver as a panacea that would boost employment, help debtors, restore prosperity and alleviate inequality. In contrast, others felt that adopting the gold standard, "sound money" and high tariffs were the keys to prosperity.[252]

The state elections of 1898 saw fierce rivalry between two "fusion" tickets. Charles Thomas ran on the silver ticket combining the Democrats, Teller Silver Republicans and Populists. Henry Wolcott ran on a gold ticket including the Republicans and "mainstream" Broad faction Silver Republicans, allied with McKinley and the national Republican Party. The Colorado election was a nationally important referendum on the future of the silver cause. The *Aspen Tribune* noted:

> *Let the intelligent voter, when he is in the privacy of the election booth consider well that the eyes of the silver men throughout the United States are watching for the result of the vote of Colorado. They expect, and have a just right to expect, that Colorado will give a tremendous majority for the silver candidates and that the Wolcott-McKinley goldbug candidates will be snowed under so deep that they won't thaw out even after the presidential election in 1900.*[253]

The Teller Silver Republicans ran an advertisement in the *Greeley Tribune* emphasizing the importance of the Colorado vote, saying:

> *There are two tickets and one issue in this campaign. The issue is gold monometallism against silver and bimetallism.... The conspiracy to destroy the Silver Republican organization has now been completed, and the combination on the gold ticket effected. The Silver Fusion ticket is headed by Mr. Charles S. Thomas as governor, and Messrs. John F. Shafroth and John C. Bell for Congress. The election of this ticket by a decisive and overwhelming majority will encourage the friends of silver everywhere, and stamp forever as false the statement now being repeated all over the East that the silver sentiment in Colorado is dead.*[254]

In opposition, Senator Wolcott denounced the silver "fusion" ticket at a Republican gathering in Colorado Springs: "He chastised the fusion ticket as a monstrosity, and the worst conglomeration ever put in the field by any party or fusion of parties." The *Colorado Daily Chieftain* noted the opposition to Thomas by Denver's *Rocky Mountain News*, using a scurrilous nickname for that newspaper: "Mr. Thomas has been denounced in terms of almost universal abuse by the Rotten Mouthing News of Denver, and by the Republican of that city." The *Chronicle News* made the assertion that Colorado was really a Republican state at heart and that

> *it was by a wave of passion and deceit that Colorado was swept into anti-Republicanism, and it is highly significant that at the present time it is no genuine party which stands in opposition to the Republicans but a chance combination of three separate parties who are only agreed in their thirst for office, in their desire for office, and their opinion that the best way to get votes is to hurrah for silver.*[255]

Most criticism of Charles Thomas focused on his career as a Democratic politician and the fact that in previous elections he had denounced the Populist Party. He also served in the Confederate army as a teenager, drawing additional ire. Thomas was lean, balding, with a neatly clipped mustache and a habitually determined, even fierce expression. After moving to Denver in 1871, he had been the city attorney in 1875–76. Active in Democratic politics, Thomas served on the Democratic National Committee for fifteen years (1884–96). He had campaigned for election as a U.S. senator in 1888 and 1895 and as governor in 1894, unsuccessfully.[256]

Ever the gadfly, former Populist governor Davis "Bloody Bridles" Waite traveled Colorado speaking against Charles Thomas, the candidate supported by his own Populist Party. At the Cripple Creek courthouse, Waite spoke to an audience of miners, saying that he was a pure "middle of the road" Populist who opposed the silver "fusion" ticket. He claimed that Thomas was an enemy to labor, called him a "shyster" lawyer and judged that Wolcott was a much better man. Waite defended the Spanish-American War, stating that it was "the most righteous war of modern times."[257]

Echoes of the Civil War also emerged in Pueblo's *Colorado Daily Chieftain* just before the election, when the newspaper printed a passionate appeal by Oliver L. Boyd to the "colored" citizens of Pueblo. He urged them to vote for Republican candidates representing the party of Abraham Lincoln,

Charles Thomas (1849–1934) was elected governor by a multi-party fusion alliance in 1898 and then won election to the U.S. Senate in 1913. *Library of Congress.*

Frederick Douglass and emancipation. Reminding them of the Democratic Party's support of the Confederacy and Jim Crow laws, he wrote:

> *There are reasons known by all why a negro should never support the democratic ticket. A democrat hates a negro 364 days every year. It is the party that advocates mob and lynch law in the south and subjects our race to all manner of degradation, and has ever been the party antagonistic to our freedom and advancement. If they had absolute control not one of our children would ever be allowed the privilege of the free school system.... Therefore I appeal to you as Afro-Americans, to remember your duty, maintaining the hand of republicanism by voting for the party of parties, for the men of men, who will in return support the administration of our beloved president, William McKinley.*[258]

The efforts of the Wolcott brothers, the Republican Party, the Broad faction "mainstream" Silver Republicans and African American voters were not enough. The Democrats, Populists and Teller Silver Republicans swept the field, with Charles Thomas being elected governor by a majority of roughly fifty thousand votes. Shafroth and Bell were also elected to

Left: Most Colorado African Americans like this miner remained loyal Republicans, grateful to the party of emancipation and Reconstruction. *Cripple Creek District Museum.*

Below: The abandoned transformer house at the Silver Lake Mine and Mill near Silverton reveals the fate of silver mining in Colorado. *Library of Congress.*

Congress by large majorities. The challenge to silver had been swept away, and Colorado would remain the Silver State.[259]

That fact was reflected in the composition of the state legislature after the election. In the State Senate there were nine Populists, nine Democrats, nine Silver Republicans, five Teller Silver Republicans, two Republicans and one National Silver Party member. In the State House of Representatives there were twenty Populists, twenty-one Democrats, sixteen Teller Silver Republicans, six Republicans and two Silver Republicans.[260]

As Colorado moved toward the beginning of a new century, free coinage of silver at sixteen to one was still the rallying cry of the determined silver apostles. To them, the return to bimetallism was a crusade, and silver beckoned like a Holy Grail of national salvation. They fully expected the silver cause to be the main issue in the 1900 presidential election with William Jennings Bryan as their candidate. Senator Henry Teller expressed their hopes and desires:

> *The contest of 1900 will not only be a contest for a righteous monetary system that will recognize the money of the constitution, but it will be for a good government, a better recognition of the rights of the people, and a return to the old doctrine of our fathers, "a government of the people, and by the people."*[261]

EPILOGUE

Free Silver as a political cause faded away during the early 1900s. Massive gold strikes in the Yukon brought large quantities of gold into the American economy. Farm prices rose even during the 1896 presidential campaign, easing the financial burden on agriculture. With additional gold available, the money supply could expand, accomplishing exactly what free silver exponents sought. Unemployment fell to a national level of about 5 percent by 1900, and much of the financial damage of the Silver Crash and Panic of 1893 was reversed. Riding high on emotions of patriotic pride after the Spanish-American War, the nation strode boldly into the twentieth century as the Panic of 1893 and the Silver Crash became just a painful memory.[262]

Silver mining in Colorado remained an important element as improved mining and smelting techniques lowered extracting and processing costs. Even during the Silver Crash in 1898, 23,502,601 ounces of silver were mined, with a value of $13,866,522. As silver deposits were exhausted, the value of Colorado's silver production declined. By 1923, 5,334,488 ounces of silver were produced, with a value of $4,374,280—less than one-third the total in 1898. The Great Depression of the 1930s and conversion to mining molybdenum, zinc and other metals important to World War II production further minimized the silver industry. Today, there is only one major producer of silver in Colorado, the Cripple Creek and Victor Gold Mine. As of 2007, about 89,000 ounces of silver emerged from that mine and 4,587 ounces from the smaller Cash Mine in Boulder

The body of a slain miner lies on the railroad tracks near Forbes, Colorado, during the "labor war" of 1914. *Library of Congress.*

County—a tiny vestige of Colorado's silver production during the great silver boom of the 1880s.[263]

Many of Colorado's **silver mining communities** never recovered from the Silver Crash. Winfield, Silver Creek, Crystal, Teller City, Carson, Animas Forks, Tin Cup and others were simply abandoned and are now ghost towns where little trace of their existence remains.

The **Leadville** population was more than 10,000 in 1890; today, about 2,600 hardy souls are willing to endure the frigid, harsh winters and isolation to enjoy the grandeur and beauty of the Rocky Mountains. Not for them are the overcrowding, crime and chaotic traffic of the "flatlands." Today, Leadville is surrounded by Superfund sites, with the EPA attempting to mitigate the toxic waste left by more than a century of gold and silver mining.[264]

Creede, the rumbustious city where there was no night, was another silver mining community of more than ten thousand residents in 1891. Now, fewer than three hundred people live there year-round.

Silverton, perhaps the most remote and isolated of all the silver towns, is growing in population, with 694 citizens now compared to 531 in 2000. The town stays alive mostly through the Durango and Silverton Narrow Gauge Railroad, whose steam-powered locomotives bring tourists along the most scenic rail route in the country. Visitors can also drive to Silverton over Red Mountain Pass, using the same route as Otto Mears's original toll road from the 1870s.[265]

Aspen, Telluride, Ouray and Durango reinvented themselves after World War II. Returning veterans of the Tenth Mountain Division created a thriving ski industry in Colorado, and the former silver mining towns became major venues for skiing, snowshoeing, ice skating and snowboarding. Music festivals and all kinds of summer recreation made these communities ripe tourist destinations. Massive new hotels, trendy restaurants, conference centers and huge second homes of the wealthy now stand where miners once struggled and died to earn three dollars a day.

Central City, Black Hawk and **Cripple Creek** became the focus of legalized gambling in Colorado when casinos were approved by the legislature only at those locations, and on the Ute Reservation, in 1991. Massive new casinos dwarf the remaining historic buildings in Black Hawk. Central City and Cripple Creek have done a better job of adapting the original structures to house casino operations. Gambling "saved" these old mining cities from dying on the vine, but it has been a mixed blessing. Summers in Central City are enlivened by the annual Central City Opera festival, with classic operas and new works being presented in the magnificent 1878 Central City Opera House.[266]

Following the failure of the Leadville strike of 1896–97, many of those who remained in the **Western Federation of Miners** were increasingly convinced that the entire capitalist system must be destroyed. Seeking the support of the American Federation of Labor, they were disillusioned by the AF of L's more conservative approach. By the time of the great mining strikes of 1903–4, the WFM's tactics had grown more violent and were matched by the mine owners to such an extent that these strikes became known as the Colorado Labor Wars. Led by "Big Bill" Haywood, the WFM employed an explosives expert named Harry Orchard to sabotage various mines, blow up a railroad platform packed with strikebreakers and assassinate former Idaho governor Frank Steunenberg. There is no doubt that the labor situation across the West had become progressively more violent—a bitter legacy of the Silver Crash.[267]

Denver recovered robustly from the depths of the Silver Crash. By 1902, the Denver population had grown to more than 170,000, and the city was strong and confident. Under Mayor Robert Speer's corrupt but imaginative leadership, Denver went from a dusty cow town to "Paris on the Platte." Broad, tree-lined boulevards, Civic Center Park, a municipal auditorium, the Denver Natural History Museum, expansion of the Denver Zoo and the magnificent Denver Mountain Parks system were among many accomplishments of the Speer administrations.

Senator Henry Teller became a Democrat as the Silver Republicans faded away after the election of 1900. He gained national fame in 1898 when he was able to affix the Teller Amendment to the congressional resolution for war with Spain. The amendment specified that the United States was not fighting for territorial gain and would not annex Cuba. Teller continued to serve the people of Colorado as their senator until his retirement in 1909. He had been a United States senator for thirty years.[268]

After his defeat in the gubernatorial race of 1894 by Albert McIntyre, former governor **Davis H. Waite** remained a pure "middle of the road" Populist and campaigned against the "fusion" Democrat/Populist/Teller Silver Republican ticket in 1898. Remaining an idealistic Populist, he spoke frequently in support of the Populist Party platform and edited a Populist newspaper, the *Nation's Crisis*. He died in November 1901 while peeling apples for Thanksgiving dinner at his home in Aspen. To the end, he crusaded for what he felt were the people's rights.[269]

Margaret Tobin Brown (1867–1932) became legendary as "unsinkable" after surviving the *Titanic* disaster. Regardless of movie portrayals, she was multilingual and cultured. *Library of Congress.*

Margaret Tobin Brown and her husband, J.J. Brown, used their newfound wealth to move to Denver and acquire a home in the fashionable Capitol Hill neighborhood. Margaret mingled in high society in Newport, New York and Europe. Learning of her grandson's illness while in Egypt in 1912, she booked passage on the first available ship to America—the *Titanic*. Lauded as a heroine of the sinking and its aftermath, Margaret described herself as "unsinkable." She was involved in progressive politics, crusading for women's suffrage, helping to found an orphanage and even contemplating a run for the State Senate. After her death in 1932, she was caricatured in Gene Fowler's highly fictionalized book *Timber Line* and later in the Broadway musical and film *The Unsinkable Molly Brown*. Though she was never known as "Molly" during her lifetime, nevertheless her home is open to the public as the Molly Brown House Museum.[270]

"Baby Doe" Tabor lived as a recluse at the Matchless Mine near Leadville, dying there in 1935. *Denver Public Library, Western History Collection Z-14609-60264.*

"Baby Doe" Tabor and her two daughters were in dire financial straits after Horace Tabor's death. On July 4, 1901, Winfield Scott Stratton returned ownership of the Matchless Mine to "Baby Doe" after a period when it was seized by creditors. Though Horace Tabor's reputed last words to "Baby Doe," "Hold on to the Matchless," are fictional, she gradually became a recluse, living in a shack at the Matchless Mine in Leadville. Local miners remembered seeing her making her way into town to buy groceries, her feet and legs wrapped in burlap sacks. She died alone in her cabin in 1935, and her frozen body was not found until several days later.[271]

William McKinley presided over a renewal of national domestic prosperity and expansion abroad. After the Spanish-American War, the United States assumed control over Puerto Rico, Guam, Wake Island, Hawaii and the Philippines. After the failure of an international conference on bimetallism, McKinley signed the bill officially putting the United States on the gold standard. Vice President Garret Hobart died in 1899, and McKinley accepted Theodore Roosevelt as his vice presidential partner in the 1900 election, though he was skeptical of Roosevelt's impetuous personality. McKinley and Roosevelt triumphed over Bryan in 1900 with their boasts of a "full dinner pail" at home and the expansion of American influence

Above: President William McKinley gives his last public address on September 5, 1901, before his assassination at the Pan-American Exposition in Buffalo, New York. *Wikipedia.*

Left: This young "powder monkey" exemplifies the spirit of Colorado miners: tough, resilient and ready for anything the world can throw at them. *Cripple Creek District Museum.*

William Jennings Bryan is ridiculed in this *Puck* cartoon for emphasizing free silver, a dead issue in the 1900 presidential campaign. *Library of Congress.*

overseas. While making an official visit to the Pan-American Exposition in Buffalo, New York, on September 5, 1901, McKinley was shot by anarchist Leon Czolgosz. He died eight days later as, some accounts say, he murmured a few lines from the hymn "Nearer My God to Thee." He is buried in a magnificent pink granite mausoleum in Canton, Ohio.[272]

William Jennings Bryan ran for president twice more. In 1900, his primary issue was anti-colonialism. Again, he was defeated by William McKinley and his vice presidential candidate, Theodore Roosevelt, now a Spanish-American War hero.

After McKinley's assassination in 1901, Roosevelt became president. In 1908, Bryan was nominated in Denver with the slogan "Shall the People Rule." Swamped by William Howard Taft, Roosevelt's handpicked successor, Bryan never again ran for public office.

Appointed as secretary of state by President Woodrow Wilson, Bryan resigned in protest when he saw the country sliding into involvement in World War I. Immensely popular on the Chautauqua circuit, Bryan helped lead crusades for women's suffrage, prohibition and banning the teaching of evolution. As part of the prosecution team in the celebrated "monkey trial" in Dayton, Tennessee, in 1925, Bryan was flummoxed and humiliated by Clarence Darrow during a cross-examination about the Bible. Bryan died shortly afterward and was vilified in a vicious diatribe by journalist H.L. Mencken. Thinly disguised as Matthew Harrison Brady, Bryan was caricatured as an ignorant buffoon in the play and film *Inherit the Wind.* Bryan and his wife, Mary, are buried in Arlington National Cemetery, commemorating his service as colonel of the Third Nebraska Volunteer Infantry during the Spanish-American War. On his simple tombstone is carved:

> *Statesman, Yet Friend to Truth: Of Soul Sincere*
> *In Action Faithful, and in Honor Clear*[273]

NOTES

Introduction

1. "Jim Corbett Dies," *San Francisco Chronicle*, February 18, 1933; Nellie Bly, "A Visit with John L. Sullivan," *New York World*, May 26, 1889.
2. "The Man Who Licked John L.," *Leadville Herald Democrat*, April 27, 1893.
3. Noel, *Guide to Colorado's Historic Places*, 267, 271–72.
4. Iverson, *Molly Brown*, 90–91.
5. Wolfenbarger, *Leadville Historic Resources Survey Plan*, 16–20, 42–44.
6. Iverson, *Molly Brown*, 101; "A Friendly Act," *Leadville Daily/Evening Chronicle*, August 16, 1889; "Caught a Slick Thief," *Leadville Daily/Evening Chronicle*, February 13, 1892; Simmons, "Labor of Love," 45–58; Edward Blair, *Leadville*, 77–80.
7. Wolfenbarger, *Leadville Historic Resources Plan*, 29, 39.
8. *Leadville Herald Democrat*, Smith's Dry Goods advertisement, May 27, 1893; "The City in Brief," May 27, 1893; "The Fearnley Block," August 19, 1893.
9. Denver Public Library, "Mine Directory—Colorado 1879–1910."
10. Blair, *Leadville*, 58–61; Iverson, *Molly Brown*, 91–93; Fell and Twitty, "Mining Industry in Colorado," Section E, 18.
11. Advertisement, *Leadville Daily/Evening Chronicle*, August 26, 1893.
12. "The Man Who Licked John L.," "In the Play," *Leadville Herald Democrat*, April 27, 1893.
13. "The Opera House," *Leadville Weekly Herald*, November 15, 1879; "Time Line of Leadville's History."

14. "Phrenology," *Leadville Daily/Evening Chronicle*, July 27, 1889; "The Great Modjeska," *Carbonate Chronicle*, July 22, 1889; "Banjo and Glee Club," *Leadville Herald Democrat*, May 19, 1892; advertisements, *Leadville Herald Democrat*, April 15, 1881, November 5, 1882, November 6, 1887, April 16, 1890, December 30, 1890, July 24, 1892, April 30, 1893; *Leadville Daily Herald*, November 4, 1882.
15. "Oscar Dear," *Leadville Daily Herald*, April 14, 1882; Wilde, "Impressions of America."
16. "The Man Who Licked John L.," *Leadville Herald Democrat*, April 27, 1893.

Chapter 1

17. Fell and Twitty, "Mining Industry in Colorado," Section E, 3–4.
18. "Song of the Stampedes," Golden *Western Mountaineer*, December 28, 1859.
19. Colorado Encyclopedia, "Colorado Gold Rush."
20. "The Restless West," *Rocky Mountain News*, March 25, 1868.
21. Gardner, "Sesquicentennial Ghost Towns," 22–27; Gardner, "Original Jefferson County," 2–4; Jurgens, "Delivering People, Goods, and Mail," 5–9.
22. "From Griffith District," *Rocky Mountain News*, May 4, 1861.
23. "From California Gulch," *Rocky Mountain News*, August 27, 1860; Dempsey and Fell, *Mining the Summit*, 16–18.
24. Fell and Twitty, "Mining Industry in Colorado," Section E, 7.
25. Gregory and Smith, *Million Dollar Highway*, 9–12.
26. "From San Juan," *Rocky Mountain News*, January 17, 1861; "From the San Juans," *Rocky Mountain News*, February 9, 1861.
27. "A Tin Cup," *Gunnison News-Democrat*, October 8, 1881; D.B. Herrick, "Hiller Once in Business Here," *Chaffee County Republican*, November 20, 1925.
28. Fell and Twitty, "Mining Industry in Colorado," Section E, 12; "From the Santa Fe Gazette," *Rocky Mountain News*, July 9, 1861.
29. Abbott, Leonard and McComb, *Colorado*, 70–76, 81–85; Josephy, *Civil War in the American West*, 296–313.
30. "Electricity," *Rocky Mountain News*, April 18, 1867; Abbott, Leonard and McComb, *Colorado*, 84.
31. Koenig, *Bryan*, 72–73, 80–86.
32. Kilgard and Narron, "Crisis Chronicles"; "How Prices Fell," *Leadville Herald Democrat*, March 13, 1892.

33. Huston, *Silver Camp Called Creede*, 45–49; "Boulder County Mines," *Boulder Daily Camera*, August 18, 1891; "The Comstock Lode," *Colorado Miner Weekly*, February 18, 1893; "Bullion," *Golden Weekly Globe*, November 24, 1877.
34. Blair, *Leadville*, 20–25.
35. Dempsey and Fell, "Mining the Summit," 26–39; Blair, *Leadville*, 58.
36. Dempsey and Fell, "Mining the Summit," 45–58.
37. "State Notes," *Colorado Transcript*, February 19, 1879.
38. "Local Matters," *Rocky Mountain News*, February 13, 1861.
39. Abbott, Leonard and McComb, *Colorado*, 123–25.
40. Gregory and Smith, *Million Dollar Highway*, 49–50, 54.
41. Ibid., 58–60; Monnett and McCarthy, *Colorado Profiles*, 83–88.
42. "The Present and Future Railroad Situation of Gunnison County and Colorado," *Gunnison News*, May 15, 1880.
43. Gregory and Smith, *Million Dollar Highway*, 20.
44. Fell and Twitty, "Mining Industry in Colorado," Section E, 21, 25.
45. Leyendecker, Bradley and Smith, *Rise of the Silver Queen*, 146–47, 227, 272–76.
46. Huston, *Silver Camp Called Creede*, 46–48.
47. Koenig, *Bryan*, 120; Blair, *Leadville*, 173–74.

Chapter 2

48. "Baring Brothers," *Colorado Daily Chieftain*, November 16, 1890.
49. "Decker, Howell & Co.," *Colorado Daily Chieftain*, December 6, 1890.
50. Koenig, *Bryan*, 93–95; "The New Tariff Law," St. Clairsville, Ohio *Belmont Chronicle*, October 16, 1890.
51. "Up Go the Prices Now," *New York Times*, October 21, 1890.
52. Brian Solender, "'Farming Don't Pay,'" 24–30, 33–38.
53. Koenig, *Bryan*, 82–84.
54. "Michigan Fires," *Castle Rock Journal*, August 17, 1887; "Drought and Forest Fires," *Colorado Transcript*, August 17, 1887; Morris, *Rise of Theodore Roosevelt*, 364–66; Bill of Rights Institute, "Cleveland and the Texas Seed Bill."
55. Koenig, *Bryan*, 82–84; Leyendecker, Bradley and Smith, *Rise of the Silver Queen*, 283–85.
56. Koenig, *Bryan*, 139; Holtz, "Folk Songs of Great Plains Homesteading," 36–45.

57. Leyendecker, Bradley and Smith, *Rise of the Silver Queen*, 272–76.
58. "Iowa—A 'Gibraltar' and a 'Vesuvius,'" *Non Partisan Leader*, March 25, 1918.
59. "Gov.-Elect Waite," *Colorado Farmer*, December 10, 1892.
60. "Suffrage and Silver," *Pagosa Springs News*, September 29, 1893; "Suffrage a Right of Citizenship," *Queen Bee*, April 26, 1893.
61. Brunk, "Reading Railroad and the Panic of 1893."
62. "With a Wild Yell," *Leadville Herald Democrat*, May 5, 1893; "Many Brokers Broke," *Aspen Weekly Times*, May 6, 1893; "He Again Goes Under," *Aspen Daily Chronicle*, May 6, 1893; "World of Finance," *Leadville Herald Democrat*, May 18, 1893.
63. Whitten, "Depression of 1893."
64. "Teller Talks Silver," "The Silver Market," *Aspen Daily Times*, June 30, 1893.
65. "He Demands Repeal," *Greeley Tribune*, August 10, 1893.
66. "No Quarter and No Compromise…," *Leadville Daily/Evening Chronicle*, August 11, 1893.
67. Colorado Encyclopedia, "Henry Teller"; "E.O. Wolcott Is No More," *Leadville Herald Democrat*, March 2, 1905.
68. "Throw Up the Game," *Aspen Daily Chronicle*, October 13, 1893.
69. "A Silver Dollar," *Leadville Herald Democrat*, August 1, 1893.
70. United States Senate, "Quorum Busting"; Congressional Quarterly Almanac, "History, Techniques of Senate Filibusters"; "Throw Up the Game," *Aspen Daily Chronicle*, October 13, 1893.
71. "The Silver Battle Ended," *Idaho Springs News*, October 27, 1893.

Chapter 3

72. "The Mines Close Down," *Leadville Herald Democrat*, June 30, 1893; Blair, *Leadville*, 174–75.
73. "Felt at Pueblo," *Greeley Tribune*, July 6, 1893.
74. "Despondent Silver Men," *Aspen Daily Chronicle*, June 27, 1893.
75. "Go to Church Today," *Aspen Daily Times*, May 13, 1894; "Free Coinage of Silver…," Silver Plume *Silver Standard*, April 22, 1893; "Listen to My Tale of Woe," *Salida Mail*, July 11, 1893.
76. "The Railroads in It," *Carbondale Avalanche*, July 9, 1893.
77. "Two Dozen Are Dead," *Colorado Daily Chieftain*, January 13, 1893; "Frightful Loss of Life," *Fort Collins Courier*, January 31, 1884.
78. Smith, *Trail of Gold and Silver*, 68–69, 199.
79. Ellis, *Life of an Ordinary Woman*, 31–32.

80. Ibid., 82.
81. "Miner's Meat Market," advertisement, *Solid Muldoon*, February 5, 1892; Backus, *Tomboy Bride*, 50–51.
82. Monnett and McCarthy, *Colorado Profiles*, 146–47; Gregory and Smith, *Million Dollar Highway*, 54, 86.
83. "Suicide," *Leadville Daily Herald*, November 10, 1880.
84. "Baseball Alexanders," *Colorado Daily Chieftain*, May 2, 1893; "Baseball Gossip," *Aspen Weekly Times*, July 27, 1895.
85. Schlereth, *Victorian America*, 212–13; Ellis, *Life of an Ordinary Woman*, 206.
86. "For Industrious Girls," *Colorado Transcript*, February 22, 1893.
87. Iverson, *Molly Brown*, 99–100, 103–4; Blair, *Leadville*, 176–77; "Greatest on Earth," *Leadville Herald Democrat*, November 26, 1893.
88. Moynihan and Watts, *Mountain Memories*, 71–73, 105–7, 147–48; Abbott, Leonard and McComb, *Colorado*, 112–15.
89. Colorado Encyclopedia, "Leadville Ice Palace"; "Some Notes from the Leadville Crystal Palace," *Chaffee County Republican*, February 26, 1896.
90. "A New Road," *Pagosa Springs News*, January 13, 1893; "Receiver Appointed," *Aspen Weekly Times*, August 5, 1893.
91. Smith, *Horace Tabor*, 40–55, 247–48; "Early Day Bonanza King Gone to His Last Home," *Leadville Herald Democrat*, April 11, 1899; Noel, *Guide to Colorado's Historic Places*, 270–72; Wyckoff, *Creating Colorado*, 113.
92. Smith, *Horace Tabor*, 227–38.
93. Horace Tabor to Elizabeth Tabor, March 11, 1894, Horace Austin Warner Tabor Manuscript Collection, MSS 614, History Colorado.
94. "On the Verge of Poverty," *Wray Weekly Times*, November 17, 1900; Smith, *Horace Tabor*, 307–8; "Tabor's Star Rising," *Rocky Ford Enterprise*, February 24, 1898.
95. Leyendecker, Bradley and Smith, *Rise of the Silver Queen*, 279–80.
96. "Grand Populist Doctrine," *Grand Valley Star*, November 4, 1893.

Chapter 4

97. Leonard and Noel, *Short History of Denver*, 9–15; Whitlock, *Distant Bugles, Distant Drums*, 73–79.
98. Leonard and Noel, *Short History of Denver*, 41–50.
99. Colorado Encyclopedia, "Five Points"; Iverson, *Molly Brown*, 107–8.
100. Noel, *Colorado Historic Places*, 90, 107-110, 128; Wyckoff, *Creating Colorado*, 113–115.

101. "Charity Organization Society," *Queen Bee*, June 4, 1890; "Homeless Children," *Rocky Mountain Sun*, June 15, 1895.
102. "A Panic in Denver," "Three More on the 19th," *West Side Citizen*, July 21, 1893.
103. "Idle Men in Colorado," *Pantagraph*, July 28, 1893.
104. "Box Car Tourists," *Leadville Herald Democrat*, July 30, 1893.
105. "Denver Is Quiet," *Leadville Herald Democrat*, July 30, 1893.
106. "The Assisted Campers in Denver," *Colorado Daily Chieftain*, August 1, 1893; "Camp Relief," *Meeker Herald*, August 5, 1893; "To Be Abandoned," *Grand Valley Star*, August 5, 1893.
107. "The Fighting Parson," *Aspen Daily Times*, October 17, 1891; Billion Graves, "Thomas A. Uzzell."
108. "Haymarket Mission," "The Maverick Woodyard," *Colorado Transcript*, January 24, 1894.
109. "Local News," *Pantagraph*, September 8, 1893.
110. Ohio History Central, "Jacob S. Coxey"; Oregon Encyclopedia, "Coxey's Army"; "Coxey's Army Over the Top at Last," *Sunday Star* (Washington D.C.), December 4, 1931.
111. "The Coxeyites," *West Side Citizen*, June 1, 1894; "Denver's Merry War," *Queen Bee*, April 4, 1894; "The Colorado Army," *Queen Bee*, May 9, 1894.
112. "A Correction," *Woman Voter*, May 31, 1894; "Important Colorado News," *Queen Bee*, June 20, 1894.
113. "Coxey's Industrial Army," *Queen Bee*, May 2, 1894.
114. "The Denver Coxeyites," *Queen Bee*, June 20, 1894; Schwantes, *Coxey's Army*, 214–19.
115. Ibid.
116. "Coxeyites Are Strapped," *West Side Citizen*, August 31, 1894.
117. "Here and There," *Woman Voter*, May 31, 1894.
118. "An A.P.A. Defense," *West Side Citizen*, February 16, 1894.
119. "Telegraphic Brevities," *Queen Bee*, June 13, 1894.
120. "Received Broken Heads," *Greeley Tribune*, September 28, 1893; "Opposed to the A.P.A.," *Aspen Daily Chronicle*, December 6, 1893; "Fanaticism Run Mad…," *Fort Collins Courier*, March 8, 1894.
121. Rove, *Triumph of William McKinley*, 175–76.
122. Smiley, *History of Denver*, 916–18.
123. "State and General," *Delta Independent*, September 25, 1895; "All Need Healing," *Rocky Ford Enterprise*, November 14, 1895; "Magic Healing," *Salida Mail*, October 25, 1895.

124. "All Need Healing," *Rocky Ford Enterprise*, October 31, 1895; "Rocky Ford to Pueblo—What Was Seen and Heard on the Round Trip," *Rocky Ford Enterprise*, November 14, 1895; "The Healer Has Gone," *Boulder Daily Camera*, November 14, 1895; "How Schlatter Perished in Mexican Mountains," *Leadville Herald Democrat*, October 23, 1909.
125. "Grander Than Expected," *Leadville Herald Democrat*, October 13, 1895; "The Colorado Pioneers," *Castle Rock Journal*, October 2, 1896; "Major Hooper Makes Great Railroad Record," *Canon City Record*, April 4, 1907.
126. "Navajo Fire Dance," *Fort Collins Courier*, September 10, 1896; "Slaves of the Silver Serpent," *Colorado Transcript*, October 23, 1895; "Pinkerton Is Robbed," *Kansas City Journal*, October 7, 1897; "Mountain and Plain Festival Amoral," *Lafayette Leader*, September 20, 1912; Emrich, *Hollywood, Colorado*, 11–13.
127. "Homeless in Denver," *Boulder Daily Camera*, June 1, 1895.

Chapter 5

128. "The Legislature of Colorado," *Colorado Daily Chieftain*, December 6, 1892; "Thousands Greet Him," *Aspen Daily Times*, July 27, 1892.
129. Colorado Encyclopedia, "Populism in Colorado"; "Telegraph News," *Durango Wage Earner*, November 28, 1901.
130. Abbott, Leonard and McComb, *Colorado*, 183–86.
131. Colorado Encyclopedia, "Women's Suffrage Movement."
132. "Colorado Marble," *Castle Rock Journal*, April 25, 1894.
133. "Vox Populi," *Colorado Transcript*, July 19, 1893.
134. "Governor Waite," *Aspen Weekly Times*, January 13, 1894; "Let Colorado Make Her Own Money," *Queen Bee*, July 11, 1894.
135. "The Governor Seeks Information," *Colorado Daily Chieftain*, January 11, 1894.
136. "Governor Waite's Plan," *Idaho Springs News*, November 24, 1893.
137. "The Mexican Dollars," *Colorado Daily Chieftain*, December 20, 1893.
138. "Kicks from the Outsiders," *Aspen Daily Chronicle*, December 8, 1893.
139. "Governor Waite's Coinage," *Colorado Daily Chieftain*, December 3, 1893.
140. "Important Colorado News," *Silver Cliff Rustler*, January 17, 1894.
141. Noel, *City and the Saloon*, 95–98; "Amazing Inside Tale of Bunco Work Is Told by 'Squealing' Gang Member," *Great Divide*, March 14, 1923.

142. Lindsey and O'Higgins, *The Beast*, 127–38.
143. "The Governor's War," "War Alarm," *Colorado Daily Chieftain*, March 15–16, 1894; "Too Many Officials," *New Castle News*, March 17, 1894.
144. "War Alarm," *Colorado Daily Chieftain*, March 16, 1894.
145. "The Dogs of War," *Leadville Herald Democrat*, March 16, 1894.
146. "The Denver See Saw," *Colorado Daily Chieftain*, March 27, 1894; "Removed at Last," *Colorado Transcript*, April 25, 1894; "Opinions," *The Avalanche*, March 27, 1894.
147. "At Cripple Creek," "A Fight Probable," *Aspen Weekly Times*, March 24, 1894.
148. "The Miners Federation," *Philipsburg [Montana] Mail*, May 25, 1893.
149. "Situation at Cripple Creek," *Aspen Weekly Times*, January 27, 1894; "Moffat's Ultimatum," *Leadville Herald Democrat*, February 2, 1894; "Strike at Cripple Creek," *Colorado Transcript*, February 7, 1894; "The Cripple Creek Strike," *Colorado Transcript*, February 14, 1894.
150. "The Seat of War," *Rocky Ford Enterprise*, March 22, 1894.
151. "Keep Away from Cripple Creek," *Aspen Weekly Times*, March 31, 1894.
152. "No Hope of Peace," *Fairplay Flume*, May 31, 1894.
153. "Cripple Creek Strike," *Aspen Weekly Times*, February 3, 1894; "Miners' Martial Law," *Colorado Weekly Chieftain*, May 31, 1894.
154. "The Deputies Arrive," "Pitched Battle for Victor," "Superintendent Shot," "Militia Called Out," *Colorado Daily Chieftain*, May 26, 1894.
155. "A Waste of Powder," *Colorado Daily Chieftain*, June 8, 1894.
156. "Militia Called Out," *Avalanche-Echo*, June 1, 1894; "Arbitration at Last," *Creede Candle*, June 8, 1894; "No Hope of Peace," *Colorado Transcript*, June 6, 1894.
157. "Peace Rules," *Colorado Daily Chieftain*, June 5, 1894.
158. Rastall, "Labor History of the Cripple Creek District," 45–47; "Perfecting Defenses," Little Falls, Minnesota *Weekly Transcript*, June 8, 1894; "Cripple Creek, the Seat of War," *Delta Independent*, June 6, 1894.
159. Rastall, "Labor History of the Cripple Creek District," 46–48.
160. "A Charter of Peace," *Colorado Weekly Chieftain*, June 14, 1894.
161. "The Western Region," *Colorado Transcript*, June 27, 1894.
162. "Over at Last," *Colorado Daily Chieftain*, June 9, 1894; "The Political Situation," *Creede Candle*, March 2, 1894; "Political and Otherwise," *Meeker Herald*, November 3, 1894; "Turpin's New War Machine…," *Colorado Weekly Chieftain*, June 7, 1894.
163. "Governor Waite Protests," *Colorado Daily Chieftain*, July 6, 1894.
164. "Waite Arrested," *Leadville Herald Democrat*, August 29, 1894.

165. "Waite Is Beaten," *Colorado Daily Chieftain*, November 7, 1894.
166. Langsdon, *Cripple Creek Strike*, 36.

Chapter 6

167. Colorado Encyclopedia, "Leadville Strike of 1896–97."
168. Ibid.
169. "Strike at Leadville," *Saguache Crescent*, June 25, 1896.
170. "Nearly All Tied Up," "At the Mines," "At the Headquarters," *Leadville Herald Democrat*, June 21, 1896.
171. "Committee Reports," *Fairplay Flume*, March 5, 1897.
172. "Society Notices," *Leadville Herald Democrat*, June 15, 1896; "The New Union," *Leadville Herald Democrat*, June 17, 1896.
173. "The Union by Its Own Witnesses," *Leadville Herald Democrat*, February 7, 1897; MSS 334, History Colorado, Leadville Strike Reports, Report to John Campion from Operator "L.C.," August 11, 1897; "Concluded Its Arduous Labors," *Leadville Herald Democrat*, February 7, 1897.
174. "Resolutions of the Miners," *Leadville Daily/Evening Chronicle*, August 20, 1896; "From Mine Managers," *Aspen Tribune*, August 14, 1896.
175. "Mines Filling with Water," *Colorado Daily Chieftain*, August 31, 1896; "The Present Strike Situation," *Leadville Daily/Evening Chronicle*, August 22, 1896; "A High Plank Fence," *Colorado Daily Chieftain*, August 22, 1896.
176. "Moral Suasion Samples," *Leadville Herald Democrat*, February 5, 1897; "Coronado Mine Working," *Loveland Register*, September 2, 1896.
177. "The Sheriff Under Arrest," *Leadville Herald Democrat*, March 8, 1896; "The Sheriff Again in Hoc," *Leadville Herald Democrat*, March 19, 1896; "Sheriff Newman on Trial," *Leadville Herald Democrat*, June 25, 1896; "Another Attempt to Persecute Sheriff," *Aspen Tribune*, May 1, 1896; "Newman Must Go to Jail," *Colorado Transcript*, December 9, 1896.
178. "Militia Ready," *Weekly Star-Times*, July 25, 1896.
179. "The Story of the Rioting," *Colorado Daily Chieftain*, September 22, 1896.
180. "Riot at Leadville," *Rocky Ford Enterprise*, September 24, 1896.
181. "Bloodshed!" "Carnage!" *Boulder Daily Camera*, September 21, 1896; "A Summary of the Damage," *Colorado Daily Chieftain*, September 22, 1896.
182. "Troops Arrive at Leadville," *Aspen Weekly Times*, September 26, 1896; "A Summary of the Damage," *Colorado Daily Chieftain*, September 22, 1896; "Still at Deadlock," *Aspen Tribune*, October 2, 1896.

183. "Showed Bitterness," *Fort Morgan Times*, October 2, 1896; "New Men Brought In," "The Associated Press Report," "Workmen Are Coming," *Colorado Daily Chieftain*, September 26, October 10, 1896.
184. "Friends of Labor," *Aspen Tribune*, October 27, 1896; "Two Funerals Yesterday," *Colorado Daily Chieftain*, September 28, 1896.
185. "Discussing Terms," *Fort Collins Courier*, January 21, 1897.
186. "Murder at Leadville," *Aspen Tribune*, December 27, 1896.
187. "Shot in Self Defense," *Leadville Herald Democrat*, January 10, 1897.
188. "To 'Do Up' the Specials," *Colorado Daily Chieftain*, January 11, 1897; "The Verdict Is Self Defense," *Leadville Herald Democrat*, January 12, 1897.
189. "Change of Venue Taken in a Number of Cases," *Aspen Daily Times*, February 28, 1897; "Court Notes," *Salida Mail*, May 11, 1897.
190. "Sheriff O'Mahoney," *Boulder Daily Camera*, December 31, 1896.
191. "Eugene V. Debs," *Leadville Herald Democrat*, January 13, 1897.
192. "The Leadville Strike," *Aspen Weekly Times*, January 23, 1897.
193. "Debs Coming to Colorado," *Fort Collins Courier*, December 31, 1896; "Miners' Mistake," *Aspen Weekly Times*, January 23, 1897; "Still Striking," *Grand Junction Daily Sentinel*, January 21, 1897.
194. "The Strike Ended," *Grand Junction Daily Sentinel*, March 10, 1897; "Troops Recalled," *Fort Morgan Times*, February 5, 1897.
195. "An Important Law," *Colorado Transcript*, April 7, 1897.
196. "They Glory in Lawlessness," *Leadville Herald Democrat*, January 3, 1897; "Echoes of the Strike," *Leadville Herald Democrat*, March 25, 1897; "Clumsy Lying," *Leadville Herald Democrat*, May 10, 1897; "A Home Built Upon Sands," *Rocky Mountain Sun*, June 12, 1897.
197. "Here Is a New Kind of Committee," *Leadville Herald Democrat*, March 25, 1897; MSS 334, History Colorado, Leadville Strike Reports, Report of Operator "L.C.," December 1, 1897.
198. "Boyce and His Mouth," *Colorado Daily Chieftain*, May 11, 1897.

Chapter 7

199. "Nominated," [Hillsboro, Ohio] *News-Herald*, June 25, 1896; "Voice of Silver Men," *Rocky Mountain Sun*, June 20, 1896.
200. Kazin, *A Godly Hero*, 54–58.
201. "Presidential Possibilities," *Boulder Daily Camera*, June 27, 1896; "Sterling Qualities Which Led to Horace Boies' Success," *New York*

Times, September 5, 1893; "Carlisle a Candidate," *Rifle Reveille*, March 20, 1896; "Three Strong National Tickets Certain," *Saguache Crescent*, April 30, 1896.

202. Rove, *Triumph of William McKinley*, 263–67.
203. Williams, *Realigning America*, 76–77.
204. Koenig, *Bryan*, 191–92.
205. Bryan and Bryan, *First Battle*, 199–206.
206. Koenig, *Bryan*, 198–99.
207. Williams, *Realigning America*, 91–92.
208. Bryan and Bryan, *First Battle*, 229–32; "The Gold Democrats," *Aspen Tribune*, July 24, 1896.
209. Koenig, *Bryan*, 213–19.
210. Rove, *Triumph of William McKinley*, 303–5; "The Situation," *Silverite-Plaindealer*, July 31, 1896.
211. "Silver Convention," *Nevada State Journal*, July 25, 1896.
212. "William McKinley, President of the United States," Britannica.com/biography/William-McKinley.
213. "William Jennings Bryan," projects.vassar.edu/1896/bryan/html; Koenig, *Bryan*, 237.
214. United States Senate, "Mark Hanna and the 1896 Election."
215. Koenig, *Bryan*, 232.
216. "Two Eagles for McKinley," *Colorado Daily Chieftain*, November 17, 1896; "To Make McKinley Pies," *Colorado Daily Chieftain*, November 20, 1896; Leech, "Front Porch Campaign," 41–42.
217. "Our War Cry," *Aspen Daily Times*, July 16, 1896.
218. "Bryan of Nebraska," *Grand Junction Weekly Star-Times*, July 18, 1896; "William Jennings Bryan," *Grand Junction Weekly Star-Times*, October 10, 1896; "One Night at Chicago," *Boulder Daily Camera*, October 2, 1896; "Courage Brother," *Rocky Mountain Sun*, October 25, 1896; "Bryan Is the Man," *Aspen Weekly Times*, July 22, 1896.
219. "Bryan's Vindictiveness," *Chaffee County Republican*, October 21, 1896; "The Friend of Labor? No," *Fort Morgan Times*, October 16, 1896.
220. "At Silver Plume," *Woman Voter*, September 20, 1894.
221. "An Empty Demagogue," *Fort Collins Courier*, July 30, 1896; "Bryan and McKinley," *Fort Collins Courier*, September 10, 1896; "Bulldozing Goes On," *Aspen Tribune*, September 16, 1896.
222. "Bryan at Fredericksburg," *Greeley Tribune*, September 4, 1896; "Cheers of Barefooted Men," *Colorado Daily Chieftain*, September 17, 1896; "Bryan in Michigan," *Silver Cliff Rustler*, October 21, 1896.

223. Koenig, *Bryan*, 231; Rove, *Triumph of William McKinley*, 333, 375; Williams, *Realigning America*, 136–37.
224. Rove, *Triumph of William McKinley*, 374; Morris, *Rise of Theodore Roosevelt*, 550–53.
225. Koenig, *Bryan*, 238; "Thirty-Five Thousand," *Aspen Tribune*, October 24, 1896.
226. "The World's Applause," *Boulder Daily Camera*, November 3, 1896.
227. Koenig, *Bryan*, 251–52; Kazin, *A Godly Hero*, 76–80; Williams, *Realigning America*, 150–51.
228. "1896 General Election Results—Colorado."
229. Lindsay, *Collected Poems*, 96–105.

Chapter 8

230. "Aspen Wants Bryan," *Aspen Tribune*, November 13, 1896; Bryan and Bryan, *First Battle*, 604.
231. "Today's News," *Boulder Daily Camera*, November 24, 1896; "Bryan in Colorado Springs," *Colorado Daily Chieftain*, November 26, 1896.
232. Kazin, *A Godly Hero*, 78–79.
233. Leech, "Front Porch Campaign," 39–40; Rove, *Triumph of William McKinley*, 377–78.
234. "Silver Forty Cents," *Colorado Daily Chieftain*, August 7, 1897; "Silver Trust a Possibility," *Silver Cliff Rustler*, December 22, 1897; "The Drop in Silver," *Fort Morgan Times*, August 27, 1897; "Japan and Silver," *Aspen Daily Times*, March 13, 1897; "Peru Bars Out Silver," *Fort Collins Courier*, May 13, 1897; Henderson, *Mining in Colorado*, 85, 93–94.
235. "Read This Carefully," *Aspen Tribune*, September 23, 1898; "Wolcott and His Fences," *Aspen Daily Times*, January 12, 1898.
236. "Towne on Hayt," *Greeley Tribune*, October 21, 1897; "State Politics," *Aspen Tribune*, August 17, 1897; "In Favor of Hayt," *Colorado Weekly Chieftain*, September 23, 1897; "The State Election," *Leadville Herald Democrat*, September 22, 1897.
237. "The Hon. John A. Gordon," *Salida Mail*, September 10, 1897; "Judge Gabbert…," *Greeley Tribune*, October 14, 1897; "Gabbert Is Elected," *Colorado Transcript*, November 10, 1897.
238. "First Gun of the State Campaign," *Colorado Daily Chieftain*, September 8, 1898; "Stained with Blood," *Aspen Tribune*, September 8, 1898.

239. "Gold, Gold, Guggenheim Gold Secured the Nomination," *Leadville Herald Democrat*, September 9, 1898; "Morality in Politics," *Leadville Herald Democrat*, September 19, 1898; "Fusion Is Possible," *Aspen Tribune*, September 9, 1898; "What They Want," *Aspen Weekly Times*, September 10, 1898; "Guggenheim Withdraws," *Daily Sentinel*, October 8, 1898.
240. Morris, *Rise of Theodore Roosevelt*, 598–610; "Latest Cuban Scheme," *Colorado Daily Chieftain*, June 22, 1898; "McKinley and Cuba," *Colorado Transcript*, June 2, 1898.
241. "The Passing of Spain," *Telluride Daily Journal*, April 9, 1898; "We Don't Like to Fight…," *Fairplay Flume*, April 8, 1898.
242. "Anybody Who Imagines…," *Glenwood Post*, April 2, 1898.
243. "Dusky-Skinned Fighting Men," *Colorado Daily Chieftain*, April 22, 1898; "A Great Send-Off," *Aspen Tribune*, April 21, 1898; "Thousand Cheered Heartily," *Colorado Weekly Chieftain*, April 21, 1898.
244. "With No Obligation…," *Colorado Transcript*, June 15, 1898; "It Looks to a Man…," *Aspen Tribune*, May 28, 1898.
245. "Colorado Guard," *Walsenburg World*, May 5, 1898.
246. Hunt, *Colorado's Volunteer Infantry*, 31–32.
247. "Colorado's Battle Song," *Salida Daily Mail*, July 1, 1898.
248. Hunt, *Colorado's Volunteer Infantry*, 31.
249. "Are Off to the War," *Fowler Tribune*, May 20, 1898.
250. "Colorado Cavalry," *Colorado Daily Chieftain*, May 31, 1898; "Judge Torrey's Boys," *Fort Morgan Times*, May 13, 1898.
251. Hunt, *Colorado's Volunteer Infantry*, 68, 92–25, 147, 207; "The Home-Coming," *Montrose Press*, September 22, 1899.
252. U.S. Industrial Commission, *Reports on the Relations and Conditions of Capital and Labor*, 41.
253. "Let the Intelligent Voter…," *Aspen Tribune*, November 6, 1898.
254. "The Teller Silver Republican Party," *Greeley Tribune*, November 3, 1898.
255. "Wolcott on Teller," *Leadville Herald Democrat*, October 9, 1898; "The People Do Not Forget," *Colorado Daily Chieftain*, November 3, 1898; "Colorado a Republican State," *Chronicle-News*, October 12, 1898.
256. Biographical Directory of the United States Congress, "Thomas, Charles Spalding."
257. "Ex-Governor Waite Talks," *San Miguel Examiner*, November 5, 1898.
258. "To Pueblo Afro-Americans," *Colorado Daily Chieftain*, November 6, 1898.

259. "Unofficial Returns," "Victory for Silver," *Craig Courier*, November 12, 1898.
260. "Its Complexion," *Montrose Enterprise*, December 10, 1898.
261. "Teller's Position," *Fort Collins Courier*, September 23, 1897.

Epilogue

262. Williams, *Realigning America*, 163–66.
263. Henderson, *Mining in Colorado*, 69; Colorado Geological Survey, *Minerals*.
264. U.S. Environmental Protection Agency, "California Gulch Leadville Colorado."
265. "Official Durango & Silverton Narrow Gauge Railroad Train."
266. World Casino Director, "Colorado Casinos and Gambling"; Smith and Clements, "Uneasy Money," 55–65.
267. Holbrook, *Rocky Mountain Revolution*, 120–29, 208–11; Lord, *Good Years*, 156–62.
268. Colorado Encyclopedia, "Henry Teller."
269. "Waite Is No More," *Leadville Herald Democrat*, November 28, 1901.
270. Iverson, *Molly Brown*.
271. "W.S. Stratton Saves the Matchless Mine," *Colorado Transcript*, June 12, 1901; "Mrs. Tabor Dies in Cabin at Mine," *Steamboat Pilot*, March 15, 1935.
272. White House, "William McKinley"; Lord, *Good Years*, 50–52, 58–62.
273. Williams, *Realigning America*, 161–63; Kazin, *A Godly Hero*, 304–6.

BIBLIOGRAPHY

Books and Periodicals

Abbott, Carl, Stephen J. Leonard and David McComb. *Colorado: A History of the Centennial State*. Boulder: Colorado Associated University Press, 1982.

Backus, Harriet. *Tomboy Bride: A Woman's Personal Account of Life in Mining Camps of the West*. Foreword by Pam Houston, afterword by Robert G. Walton. Boulder, CO: Pruett Publishing, 1977.

Blair, Edward. *Leadville, Colorado's Magic City*. Boulder, CO: Pruett Publishing Company, 1980.

Bryan, William Jennings, and Mary Bryan. *The First Battle: A Story of the Campaign of 1896*. Chicago: W.R. Conkey Company, 1896.

Byrd, Robert. *The Senate 1789–1989: Addresses on the History of the United States Senate*. Vol. 2, Bicentennial Edition. Edited by Wendy Wolff. Washington, D.C.: U.S. Senate Printing Office, U.S. Government Printing Office, 1988.

Colorado State Federation of Labor. *Official Labor Directory and Manual of the Colorado State Federation of Labor*. Denver: Colorado State Federation of Labor, 1907.

DeLorme, Roland B. "Turn-of-the-Century Denver: An Invitation to Reform." *Colorado Magazine* 45, no. 1 (1968): 1–41.

Dempsey, Stanley, and James E. Fell Jr. *Mining the Summit: Colorado's Ten Mile District, 1860–1960*. Norman: University of Oklahoma Press, 1986.

Ellis, Anne. *The Life of an Ordinary Woman*. Foreword by Carol Bly, introduction by Lucy Fitch Perkins. Boston: Houghton Mifflin Company, 1929. Renewed 1957, foreword 1990.

Emrich, David. *Hollywood, Colorado: The Selig Polyscope Company and the Colorado Motion Picture Company*. Lakewood, CO: Post Modern Company, 1997.

Gardner, Richard. "Original Jefferson County." *Historically Jeffco* 23, no. 31 (2010): 2–4.

———. "Sesquicentennial Ghost Towns of Jefferson County." *Historically Jeffco* 22, no. 30 (2009): 22–27.

Greenberg, Davis. *Republic of Spin: An Inside History of the American Presidency*. New York: W.W. Norton and Company, 2017.

Gregory, Martin, and P. David Smith. Photography by Kathleen Norris Cook. *The Million Dollar Highway: Colorado's Most Spectacular Seventy Miles*. Ouray, CO: Wayfinder Press, 1986.

Henderson, Charles W. *Mining in Colorado: A History of Discovery, Development and Production*. Washington, D.C.: Government Printing Office, 1923.

Hewitt, William L. "The Election of 1896: Two Factions Square Off." *Colorado Magazine* 54 (Winter 1977): 44–57.

Hill, Emma S. *A Dangerous Crossing and What Happened on the Other Side*. Denver: Press of the Smith-Brooks Company, 1914.

Holbrook, Stewart H. *The Rocky Mountain Revolution*. New York: Henry Holt and Company, 1956.

Holtz, Dan. "The Folk Songs of Great Plains Homesteading: Anthems, Laments, and Political Songs." *Nebraska History* 94 (2013): 36–45.

Hunt, Geoffrey R. *Colorado's Volunteer Infantry in the Philippine Wars 1898–1899*. Albuquerque: University of New Mexico Press, 2006.

Huston, Richard C. *A Silver Camp Called Creede: A Century of Mining*. Under the Auspices of the Creede Historical Society. Montrose, CO: Western Reflections Publishing Company, 2005.

Iverson, Kristen. Foreword by Muffet Brown. *Molly Brown: Unraveling the Myth: The True Life Story of the Titanic's Most Famous Survivor*. Boulder, CO: Johnson Books, 1999.

Josephy, Alvin W., Jr. *The Civil War in the American West*. New York: Alfred A. Knopf, 1992.

Jurgens, Pat. "Delivering People, Goods, and Mail in the Mountains of Jefferson County." *Historically Jeffco* 23, no. 31 (2010): 5–9.

Kazin, Michael. *A Godly Hero: The Life of William Jennings Bryan*. New York: Alfred A. Knopf, 2006.

Kellie, Luna. *Memoirs of a Prairie Populist*. Iowa City: University of Iowa Press, 1992.

Koenig, Louis W. *Bryan: A Political Biography of William Jennings Bryan*. New York: Capricorn Books, G.P. Putnam's Sons, 1975.

Langsdon, Emma Florence. *The Cripple Creek Strike: A History of Industrial Wars in Colorado, 1903–4–5, Being a Complete and Concise History of the Effects of Organized Capital to Crush Unionism*. Denver: Great Western Publishing Company, 1905.

Leech, Margaret. "The Front Porch Campaign." *American Heritage* XI, no. 1 (December 1959): 32–45.

Leonard, Stephen J., and Thomas J. Noel. *A Short History of Denver*. Reno: University of Nevada Press, 2016.

Leonard, Stephen J., Thomas J. Noel and Donald L. Walker. *Honest John Shafroth: A Colorado Reformer*. Denver: Colorado Historical Society, 2003.

Leyendecker, Liston E., Christine A. Bradley and Duane Smith. *The Rise of the Silver Queen: Georgetown, Colorado, 1859–1896*. Boulder: University Press of Colorado, 2005.

Lindsay, Vachel. *Collected Poems*. New York: MacMillan Company, 1925.

Lindsey, Ben B., with Harvey J. O'Higgins. Foreword by Stephen J. Leonard. *The Beast*. Boulder: University Press of Colorado, 2009.

Lord, Walter. *The Good Years: From 1900 to the First World War*. New York: Harper & Brothers, Publishers, 1960.

Monnett, John H., and Michael McCarthy. *Colorado Profiles: Men and Women Who Shaped the Centennial State*. Evergreen, CO: Cordillera Press, Inc., 1987.

Morris, Edmund. *The Rise of Theodore Roosevelt*. New York: Ballantine Books, 1979.

Moynihan, Betty, and Helen E. Watts, eds. *Mountain Memories: From Coffee Pot Hill to Medlen Town: A History of the Inter-Canyon Area of Southwest Jefferson County*. Lakewood, CO: Limited Publications, 1981.

Noel, Thomas J. *The City and the Saloon: Denver, 1858–1916*. Niwot: University Press of Colorado, 1996.

———. *Guide to Colorado's Historic Places*. Englewood, CO: Westcliffe Publishers, 2006.

Rastall, Benjamin McKie. "The Labor History of the Cripple Creek District: A Study in Industrial Evolution." PhD diss., University of Wisconsin, 1904.

Rove, Karl. *The Triumph of William McKinley: Why the Election of 1896 Still Matters*. New York: Simon and Schuster, 2015.

Schlereth, Thomas J. *Victorian America: Transformations in Everyday Life*. New York: Harper Perennial, 1991.

Schwantes, Carlos A. *Coxey's Army: An American Odyssey*. Moscow: University of Idaho Press, 1994.

Simmons, Darby. "Labor of Love: Prostitution and Civic Engagement in Leadville, Colorado, 1870–1915." Honor's thesis, History Department, University of Colorado–Boulder, April 11, 2017.

Smiley, Jerome. *History of Denver: With Outlines of the Earlier History of the Rocky Mountain Country.* Denver: J.H. Williamson, 1901.

Smith, Duane A. *Horace Tabor, His Life and the Legend.* Niwot: University Press of Colorado, 1989.

———. *The Trail of Gold and Silver: Mining in Colorado, 1859–2009.* Boulder: University of Colorado Press, 2009.

Smith, Duane A., and Eric L. Clements. "Uneasy Money: The High-Stakes Consequences of Low-Stakes Gambling for Black Hawk, Central City and Cripple Creek, Colorado." *Mining History Journal* 4 (1997): 55–65.

Solender, Brian. "'Farming Don't Pay': The Anatomy of the 19th Century Western Farm Mortgage Industry." Senior thesis, Department of History, Columbia University, April 5, 2017.

U.S. Industrial Commission. *Report of the Industrial Commission on the Relations and Conditions of Capital and Labor Employed in the Mining Industry Including Testimony, Review of Evidence, and Topical Digest.* Vol. 12 of the Commission's Reports, Washington, D.C.: Government Printing Office, 1901.

Whitlock, Flint. *Distant Bugles, Distant Drums: The Union Response to the Confederate Invasion of New Mexico.* Boulder: University Press of Colorado, 2006.

Williams, R. Hal. *Realigning America: McKinley, Bryan, and the Remarkable Election of 1896.* Lawrence: University Press of Kansas, 2010.

Wolfenbarger, Deon. *Leadville Historic Resources Survey Plan.* Three Gables Preservation, prepared for Leadville Historic Preservation Commission, 2015.

Wyckoff, William. *Creating Colorado: The Making of a Western American Landscape, 1860–1940.* New York: Yale University Press, 1999.

Digital Sources

Billion Graves. "Thomas A. Uzzell—He Made Religion Work." billiongraves.com/grave/Thomas-A-Uzzell/13676650.

Bill of Rights Institute. "Cleveland and the Texas Seed Bill." billofrightsinstitute.org/e-lessons/cleveland-and-the-texas-seed-bill.

Biographical Directory of the United States Congress 1774–Present. "Thomas, Charles Spalding." bioguideretro.congress.gov/Home/MemberDetails.

Britannica. "William McKinley, President of the United States." Britannica.com/biography/William-McKinley.

Brunk, Daniel H. "The Reading Railroad and the Panic of 1893," The Economic Historian. economic-historian.com/2020/08/the-reading-railroad-and-the-panic-of-1893/.

Colorado Encyclopedia. "Colorado Gold Rush." coloradoencyclopedia.org/article/colorado-gold-rush.

———. "Five Points." coloradoencyclopedia.org/article/five-points.

———. "Henry Teller." coloradoencyclopedia.org/article/henry-teller.

———. "Leadville Strike of 1896–97." coloradoencyclopedia.org/article/Leadville-strike-1896-97.

———. "Populism in Colorado." coloradoencyclopedia.org/article/populism-colorado.

———. "Women's Suffrage Movement." coloradoencyclopedia.org/articles/womens-suffrage-movement.

Colorado Geological Survey. "Minerals." coloradogeologicalsurvey.org/minerals/metals.

Congressional Quarterly Almanac. "History, Techniques of Senate Filibusters." 1957, 13th ed. library.cgpress.com/cqalmanac/cqo157-1345249.

Denver Public Library. "Mine Directory—Colorado 1879–1910." history.denverlibrary.org/sites/history/files/CO1879-1910mineDirect.pdf.

"1896 General Election Results—Colorado." uselectionatlas.org/RESULTS/state.

Fell, James E., and Eric Twitty. National Register of Historic Places Multiple Property Documentation Form. "The Mining Industry in Colorado." June 30, 2006 (Revised July 27, 2008). historycolorado.org/sites/default/files/media/document/2017/651.pdf.

"Globeville-Elyria-Swansea." historycolorado.org/globeville-elyria-swansea.

Kilgard, Thomas, and James Narron. "Crisis Chronicles: The Long Depression and the Panic of 1873." Federal Reserve Bank of New York, Liberty Street Economics. libertystreeteconomics.newyorkfed.org/2016/02/crisis-chronicles-the-long-depression-and-the-panic-of-1973.html.

Kreitner, Ray. "Money in the 1890s: The Circulation of Politics, Economics, and Law." law.uci/lawreview/Vol1No3Articles/kreitner.pdf. 975-1013.

"Official Durango & Silverton Narrow Gauge Railroad Train." www.durangotrain.com.

Ohio History Central. "Jacob S. Coxey." ohiohistorycentral.org/Jacob-S-Coxey.

Oregon Encyclopedia. "Coxey's Army." oregonencyclopedia.org/articles/coxey-s-army/#.YIXNSJBKhPY.

"A Time Line of Leadville's History." Leadville.com/a-time-line-of-leadvilles-history.

United States Senate. "Mark Hanna and the 1896 Election." senate.gov/artandhistory/history/minute/Hanna_1896Election.htm.

———. "Quorum Busting." senate.gov/about/powers-procedures/rules-procedures/quorum-busting.htm.

U.S. Environmental Protection Agency. "California Gulch Leadville CO." cumulis.epa.gov/supercpad/cursites/csitinfo.cfm?id=0801478.

The White House. "William McKinley." whitehouse.gov/?s=william+mckinley.

Whitten, David O. "The Depression of 1893." EH.net. eh.net/encyclopedia/the-depression-of-1893.

Wilde, Oscar. "Impressions of America." wilde-online.info/impressions-of-america-page3.html.

"William Jennings Bryan." projects.vassar.edu/1896/bryan.html.

World Casino Director. "Colorado Casinos and Gambling." worldcasinodirectory.com/Colorado.

Manuscript Sources

Horace Austin Warner Tabor Collection, MSS 614, History Colorado.

Ibex Mining Company Collection, MSS 71, History Colorado.

Leadville Strike Reports, MSS 334, History Colorado.

Oral History Transcriptions, History Colorado, Henry Weber PAM.346.22.

———, John Matthews PAM.346.28.

———, Martha R. Mahon and John E. Paul PAM.346.15.

———, Thomas J. Doyle PAM.346.22.

———, Tobias Mattox PAM.343.20.

INDEX

E

G

H

I

J

L

M

N

O

P

R

ABOUT THE AUTHOR

John Steinle, a native of Hamilton, Ohio, earned his Bachelor of Fine Arts degree from the University of Dayton in 1972. After several years as a commercial artist, he decided to pursue a career in history. After receiving a master's degree in museum and archival management from Wright State University, Steinle worked as a curator and archivist at the Cincinnati Art Museum and the Cincinnati Historical Society. He then served as director at several Ohio museums before migrating to Colorado in 1992. In 1994, Steinle became the administrator of the Hiwan Homestead Museum in Evergreen, Colorado, working for Jefferson County Open Space. He was promoted to history education supervisor and region supervisor for the Bear Creek Region. He retired in 2016.

Steinle is the co-author with Richard Scamyhorn of *Stockades in the Wilderness: The Frontier Forts and Settlements of Southwestern Ohio, 1788–1795*, plus *Evergreen* and *Conifer* for Arcadia Publishing. Since 1975, Steinle has been a member of many different living history and reenactment groups such as the First Colorado Volunteers Civil War unit and the American Military Living History Association. He and his wife, Mary, live in Arvada, Colorado.